To Margo,
my 'partner skipper' for life,
and to
the wives and sweethearts of
fishermen everywhere

All author's royalties from the sale of this book will
be donated directly to the
Royal National Lifeboat Institution

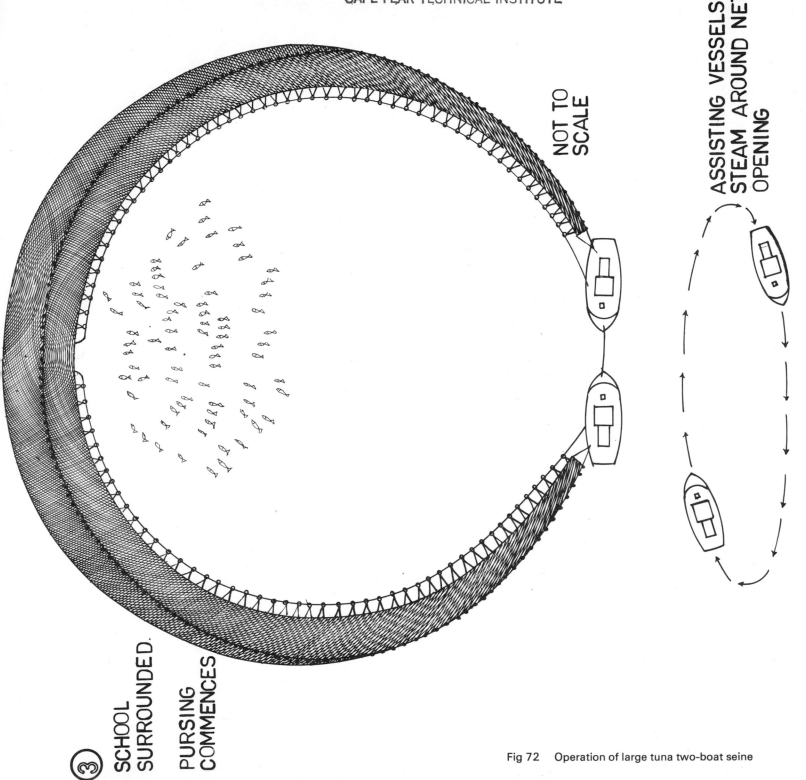

NOT TO SCALE

ASSISTING VESSELS
STEAM AROUND NET
OPENING

③ SCHOOL SURROUNDED.
PURSING COMMENCES

Fig 72   Operation of large tuna two-boat seine

David Thomson

Pair Trawling and Pair Seining

—The Technology of Two-boat Fishing

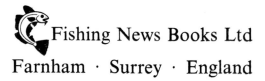

Fishing News Books Ltd
Farnham · Surrey · England

ISBN 0 85238 087 9

British Library CIP Data
Thomson, David Baikie
    Pair trawling and pair seining.
    1. Fisheries
    I. Title
    639′.22   SH344

Printed in Great Britain by
The Whitefriars Press Ltd
London and
Tonbridge, Kent

Contents

List of Figures

In each of the ten chapters describing various nets and their method of operation are drawings consecutively numbered as figures. Also included in the chapters are photographs illustrative of vessels, gears and operating practices. These are listed as plates and also numbered consecutively. These numberings are not linked to the text by direct numerical sequence but their captions or legends provide the running story.

List of Plates

Plates

Foreword

Sir William S. Duthie wrote in the foreword to Mr. Thomson's best seller *The Seine Net,* that he hoped the author would not permit his pen to run dry until he had dealt just as adequately with all other proved, permitted and worthwhile ways and means of taking fish from the sea.

Pair fishing is a proven method of catching fish, permitted subject to regulations as to the size of mesh of the nets, and most certainly worthwhile from the economic results it produces.

Since its publication eight years ago, David Thomson's first book – *The Seine Net: Its Origin, Evolution, and Use* – has justly received, for its practical and academic content alike, international recognition as the authoritative treatise on this method of fishing.

For his latest work – *Pair Trawling and Pair Seining: The Technology of Two-Boat Fishing* – I am sure no less acclaim will accrue to the reputation of a writer who brings to his subject not only a wealth of practical experience and theoretical understanding but also presents in a happy lucidity of style the importance of the inter-relationship of both of these attributes in the making of the successful fisherman.

In Scotland, the comment – 'I knew his father' – is, regrettably, most usually offered in deprecation of an individual's abilities and achievements. But many of us from the Moray Firth and North East fishing coasts of Scotland did know his father, Skipper Jimmy Thomson, and his uncles and cousins – an outstanding family of men who were pioneers, innovators, note-takers and instructors. The influence of this family is widespread not only in Scotland, but they pioneered modern fishing methods in Northern Ireland and Eire. We rejoiced to recognise in David, as a young man, early evidence not only of his lively inheritance of this notable background but also the additional clear promise of a potential to contribute from his generation a further lustre to an accumulated legacy of talent built up over generations of his family.

What an extraordinary fulfilment of that early promise he has achieved, and how proud we are that this illustrious son of Lossiemouth in his maturity is now a universally respected figure in the fishing nations of the world.

In conclusion I wish this volume well. It is not only a textbook, but an historical record which will be used as a reference for our times in the future. 'Some few books,' says Francis Bacon, 'have to be read wholly and with diligence and attention.' This in my view is surely one of them.

Dr. W. J. LYON-DEAN, OBE

Acknowledgements

This publication was compiled in an attempt to answer an increasing demand for information on pair fishing. The writer tried to put all available and useful information on the subject in a simple concise form, using illustrations rather than words wherever possible. Any mistakes or errors in the technical specifications should be blamed on the author and not on his sources. Any opinions expressed are those of the writer alone, and do not necessarily reflect the thinking of those mentioned hereafter. Much gratitude is due to those fishermen, writers, technologists and manufacturers who contributed directly or indirectly to the pool of information on this subject. It is hoped that readers, particularly fishermen, will contribute further data from their own experience, from time to time, through the fisheries press. This is the only sure way to keep the fishing communities abreast of current technology.

Much credit is due to personal friends in the fishing fraternity for a wealth of factual data and advice. Chief among these is Mr. Leslie Innes of Fraserburgh, one of the finest all-round fishing technologists in the world. Many commercial fishermen have yielded valuable information, including George Watt (MB *Kallista*), Willie Campbell (MB *Ajax*), John Thomson (MB *Horizon*), and many others who can only be included in their well known fishing families such as the Chambers of Annalong, the Munros of Ayrshire, the McCalligs of Killybegs, the Mansons of Mallaig, the Buchans of Peterhead, the O'Driscolls of Castletownbere, and many more. Skipper Norman Townsend of Grimsby Technical College, and Captain Murdo MacInnes of Tom Sleights Ltd. gave useful data on the cod pair trawling. The writer's father Jimmy Thomson, and his Canadian colleagues, Jack Rycroft and Wes Johnson submitted reports and data on pair operations in Canada. From America, information was provided by Robert Taber and Dr. Thomas Meade of Rhode Island, and R. Barry Fisher of Oregon. The report and illustrations by Mr. Tumilty of the White Fish Authority were most informative. No work on fishing gear can ignore the excellent books by John Garner, nor Dr. Joachim Scharfe's comprehensive works on midwater trawling. Much help was obtained from the late F. M. Davis's book, and from the many FAO Fisheries Books. Individuals from other countries to whom the writer is indebted are: Mr. S. Okonski, Poland; Mr. C. Nedelec, France; Mr. Arthur Reynolds, Ireland; Mr. Jose Almanar Sansaloni, Spain; Mr. E. Kvaran, Iceland; Dr. N. Nomura, Thailand; Mr. Bong Soo Kahng, Korea; Mr. Patricio Dinglasan, Philippines; and Dr. Y. Iitaka, Japan. To these, and many more too numerous to mention, the author extends sincere thanks and appreciation. Thanks are also due to the staff and correspondents of *Fishing News* for material submitted. Jane Strachan, Tom Wood and Patricia O'Driscoll were helpful in this respect.

Finally the author would like to thank Dr. W. J. Lyon-Dean for generously writing the foreword – a foreword of which the author feels most unworthy.

DAVID THOMSON

Introduction

Fishing with Two Boats

The concept of using two or more vessels in a fishing operation is not new. Fishermen on African lakes worked drive-in gill nets this way, from their dug-out canoes. Seine fishermen in many countries could work their large nets no other way before powerful engines and winches came on the scene. Two thousand years ago the fishermen on Lake Galilee 'beckoned to their partners' to assist in the hauling of their ring net. But if the concept itself is not new, its application in modern fisheries is rather novel. The trend in the past 50 years has been for fishing vessels to be equipped and automated as far as possible to make them independent units at sea in the search, capture and transport of fish. This is particularly obvious in the development of large deep sea fishing vessels which, though they may operate as a fleet, do not rely on each other at all during the fishing operations.

Inshore fishing craft have also become extremely well equipped units. Some modern 50-footers have more electronic equipment on the bridge than many an ocean-going merchant vessel. The term 'inshore' really does not apply to these craft any more (if it ever did) as they may fish hundreds of miles away from their home port. It is amongst these privately owned near-water and middle-distance vessels that one finds the greatest variety in types of fishing gear and techniques. Deep sea vessels by contrast tend to use a standard type of gear with little variation in rigging or operation. Two-boat fishing, then, was developed by the smaller vessel fishermen who were more versatile in their fishing skills, and, being the owners of their craft, were free to pursue their own ideas which a deep sea man could not do without company permission.

Two-boat fishing proved in many cases to be more successful and more economical than single-boat operation. This often surprises outsiders who think,

naturally, that it must cost more to operate two vessels. They assume that one net catches the same amount of fish as another. In actual fact successful pair fishing can be far more productive than one-boat operation in terms of investment, man-power, or any other viewpoint. But the pros and cons need not be argued on paper, the facts speak for themselves. One can compare the results of 60-foot pair trawlers in Grimsby with those of the deep sea vessels from the same port. Ships twice the size, with double the number of crew and involving a far larger investment have a similar daily fishing average. That is for bottom trawling. Midwater trawling is now successfully carried on in Europe, Scandinavia and North America. Despite considerable efforts on the part of research institutions, single-boat midwater trawling is limited chiefly to vessels of over 200 tons (or over 100 feet in length). With a few exceptions, the hundreds of boats fishing midwater trawls commercially, under those sizes, use the two-boat trawling techniques. On the other hand some pair fishing methods have been displaced by single boat techniques, notably purse seining.

Many people speak of the psychological problems of pair fishing as fishing skippers tend to be very independent types. The difficulties are rather exaggerated. Usually one finds partnerships to be formed of one dominant and one less forceful skipper, but many strong individualists have succeeded in working well together. One might compare the situation to that of a pair of musical composers, a sports team or a partnership of entertainers. The strains of those professional relationships are probably much greater than those of the pair fishermen.

Two-boat fishing prospects look even better now, given the current high cost of fuel which is making economy of operation a major factor. The success of small Canadian vessels using the pair seine techniques demonstrates the potential of lightly powered units. In the early days of trawling, more power meant more fish. Today more power means excessive running costs. By adopting new kinds of gear and different fishing techniques, skippers are learning how to catch more fish with less power.

This book attempts to describe the chief methods of pair fishing. Some attention is paid to the historical background and development of the techniques, but the main focus is on successful modern methods of fishing with two boats.

Some readers may be surprised to learn how widespread are two-boat activities. Modern fishing vessels in Europe, Asia and the Americas engage in pair fishing techniques of various kinds. Two-boat fishing also encompasses a range of craft from small open skiffs powered by outboard engines, to huge deep sea trawlers of 2,000 hp or more. The nets used vary from the tiny freshwater seines and trawls to the largest nets in the world. The biggest trawls and purse seines in existence are operated by pair fishing craft. The old proverb says that 'two are better than one' and in many cases it can be true of fishing vessels. Interest in pair fishing is on the increase following the success of recent operations in Europe and North America. For those interested in fishing with two boats, it is hoped that this book will help to clarify the different kinds of gear and techniques in use.

Plate 1   A traditional pareja trawler in operation off the West of Ireland in the 1950's   *D M Baikie*

Plate 2   Another pareja trawler of same type and era
D M Baikie

15

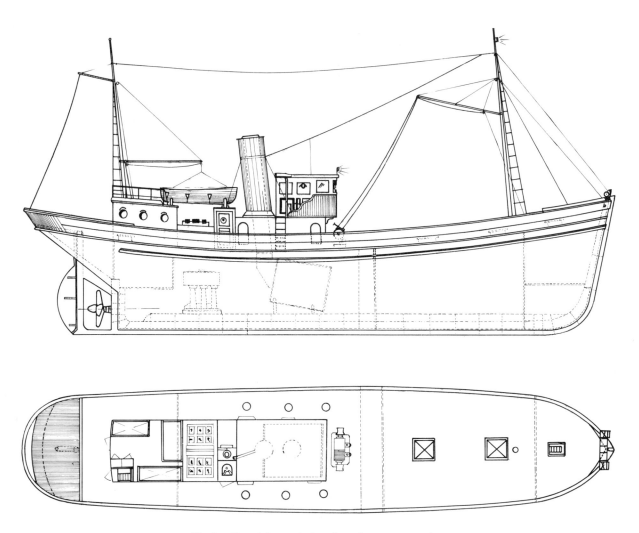

Fig 1   Spanish pareja trawler: steam powered

Fishing Boats of the World 3

1 Pareja Trawling

The first major pair-trawl fishery in the Atlantic was developed by Spanish fishermen. They were working the east Atlantic shelf chiefly for one species – hake. These much sought after fish *(Merluccius merluccius)* were found on the sea bed from north Africa to west Scotland. Preferring the deeper waters they were concentrated mostly around the 100 fathoms line but were often as deep as 200 fathoms or more. It was possible to catch hake with an otter trawl but to operate one in such depths vessels needed a lot of power. Otter trawls were also prone to fouling as they were sinking through the cross currents in deep water. So, for reasons of ease of operation, Spanish trawlermen developed the two-boat system and it proved to be remarkably effective. Not only was it a simple type of gear compared to an otter trawl; it was much more effective in catching hake. The pareja fishermen also discovered that by using long cables or bridles with which to 'sweep' the sea bed on either side ahead of the net, it was possible to catch even more fish.

By 1925 Spanish pair trawlers were working as far north as Ireland. Since then the pareja boats have been a common sight from Bantry Bay to the Hebrides. The hake fishery reached a peak in the 1930s and again in the late 1940s after the Second World War, but it declined quickly thereafter owing to too much fishing pressure on one species. While the hake fishing lasted, the pareja trawlers were the 'kingpins'. Pair after pair of vessels could be seen towing up the edge of the shelf. It was a standing joke amongst British fishermen that they shot their nets off Vigo and did not haul them till they reached the Butt of Lewis!

With the decline of hake catches in the 1950s the Spanish fleets turned their attention to cod, and began to look further afield for the fish. They found the pair trawl, with some modifications, suitable for cod fishing. Distant-water vessels were designed and built to fish on the west Atlantic shelf, on George's Bank, the Grand Banks, and the grounds off Labrador and Greenland. They developed satisfactory methods of salting the fish to preserve them on such long voyages and this is still done today, though some pareja boats are equipped with refrigeration. Some vessels still work the Irish grounds, catching mainly coalfish, Norway lobster, cod and – as a reminder of old times – a few hake.

Pareja Net Design

The pareja net was originally made of hemp, cotton and manila, with glass floats to withstand the great pressure at over 100 fathoms. It looked, and was constructed, like a huge Danish seine, *ie* of two side pieces instead of an upper and a lower section as in the case of an otter trawl. The headline measured 200-300 feet or more and it carried as many as 200 floats. The groundrope was made of tarred manila with short lengths of chain attached for weight. The cod end was fitted with a zipper line or removable lacing to allow brailing of fish when catches were large. Dan lenos or bridle poles were attached to the wing end and to these were shackled the long cables made of combination rope. The cables were made up in 50 fathom lengths and in deep water, as many as five or six lengths might be used. In addition to the long cables, a warp length of four to six times the depth was paid out.

The net design comprised long wing sections of 5- or 6-inch mesh size, and relatively short body panels. At the 'roof' and 'belly' of the bag were fitted frame strips of slightly larger mesh than the bag-side sections. Near the cod end, the lower bag had a fish-tail shape of the same larger mesh size, and the cod end itself was wide at the tail. Mature hake are large fish, hence the big meshes, but as the stock declined and fish became smaller the pareja mesh sizes were reduced. Despite the long headline, the net had an overhang of only 6 or 9 feet. On later models this was increased. However, for hake fishing, a high net mouth opening was more important than overhang. The whole rig was relatively light. It skimmed over the sea bed more like a Danish seine than a deep sea otter trawl.

Pareja Vessels

The first pareja trawlers were steam-powered wooden boats of around 80 feet. Larger steel vessels were built and diesel engines later replaced steam. The east Atlantic pareja boats are mostly around 80-120 feet in size but larger ones fish the west Atlantic grounds. The vessels are fairly conventional in appearance. Apart from the absence of gallows, they could be mistaken for ordinary side trawlers. A close inspection, however, reveals a hauling roller on the stem with sheaves leading to the winch. Aft of the casing there is a towing block fixed amidships and a large platform to accommodate the net. Pareja trawlers shoot the net over the stern and haul the warps and cables in over the stem (Plates 1 and 2).

The early steam-powered pareja boats looked not

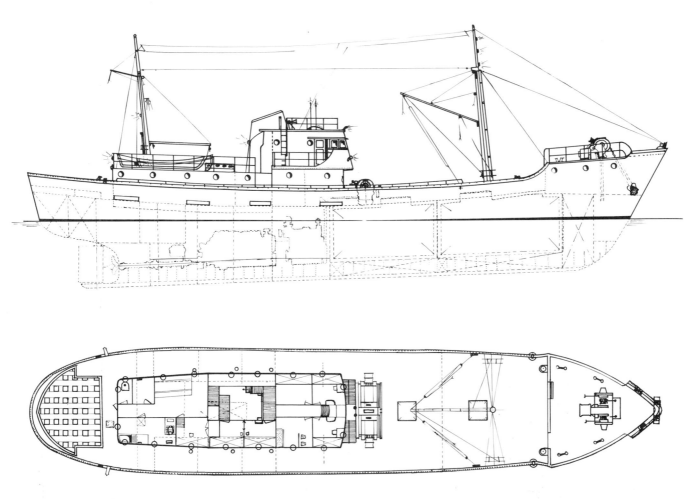

Fig 2   Pareja trawler: diesel type

Fishing Boats of the World 1

unlike a British steam drifter, except that the foremast was never stepped or lowered. A typical vessel measured 82 feet overall, with a beam of 18.2 feet and draft of 9.9 feet. It was powered by a triple expansion engine with a Scottish boiler, generating 120 hp. The 237 gross ton vessel had a service speed of 8.5 knots and could carry up to 40 tons of fish and ice. Its 45 tons of coal and 19 tons of water made it possible to spend 15 days at sea under full power. As these boats often anchored or drifted at night, trip duration was sometimes even longer than this.

A typical modern pareja boat measures 116 feet overall with a beam of 22.5 feet and draft of 12.8 feet. It has a 450 hp diesel engine and can cruise at 12 knots for up to 9,000 miles. With a gross tonnage of around 350 the boat can carry 110 tons of fish and ice. The deck layout is little different from that on the old steam-powered vessels (Fig. 2).

Fishing Techniques

The Spanish pair trawl is towed at relatively slow speeds for long periods. One haul may last from 5 to 7 hours. In the winter time only one tow is made per day, and in the summer time up to three hauls may

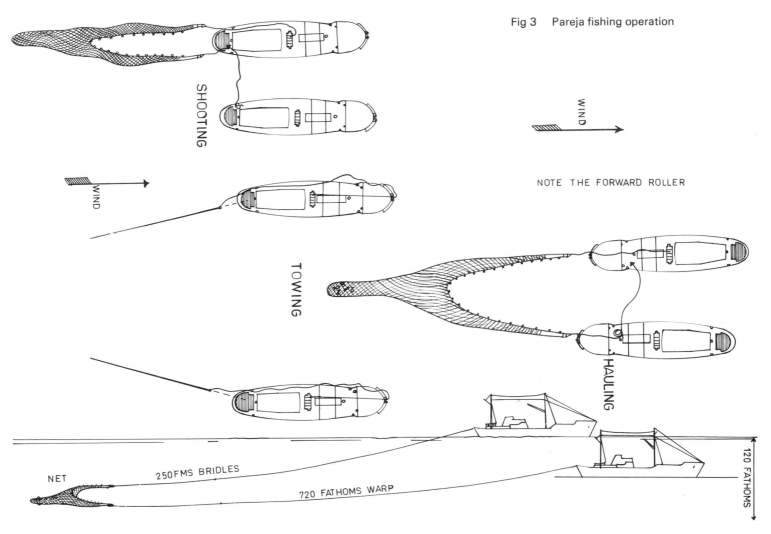

Fig 3   Pareja fishing operation

SHOOTING

WIND

NOTE THE FORWARD ROLLER

WIND

TOWING

HAULING

NET   250 FMS BRIDLES   720 FATHOMS WARP   120 FATHOMS

19

be made. Hake fishing is best during the brightest daylight hours and few are ever caught at night. Owing to the long time it takes to haul in the mile-long warp it is uneconomical to make short tows.

Setting commences with one vessel putting the net out over the stern. The second boat approaches and throws over a line to which is attached the end of its ground cable. This is connected to the net bridle and at an appropriate signal, both boats pay out the agreed length of cables and warp. In the past they communicated with each other by means of flags or whistle signals. Today radio-telephone is used. Once the warp is all paid out, towing commences with the ships remaining about half a mile apart.

The warp is connected to the towing point aft by means of a short chain shackled into a connecting link in the warp. The bight of warp between there and the winch is pulled out till there is enough slack to go around the forward sheaves and rollers ready for hauling. Hauling commences at a given signal and both ships 'knock out' the warp aft and turn to face the gear with the warp leading in over the stem roller. The warps are pulled in at full speed until the net is reached, then one ship releases its cable and

Fig 4   Traditional pareja trawl net

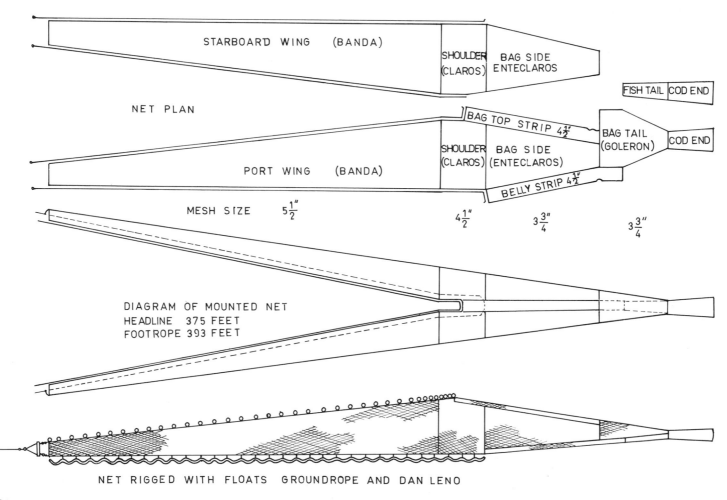

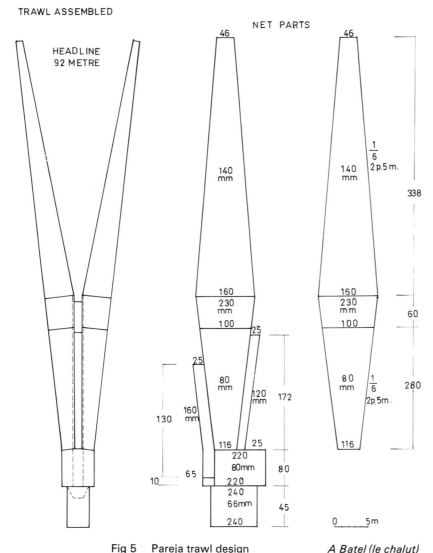

TRAWL ASSEMBLED

HEADLINE
92 METRE

NET PARTS

Fig 5   Pareja trawl design

A Batel (le chalut)

throws a line to the other, attached to the dan leno. The other boat proceeds to haul the net in over the side.

Anyone who has ever seen a bag of hake will appreciate the excitement on board when a full cod end of these large buoyant fish surfaces like a whale. In the great days of the hake fishing it was not unknown for nets to surface before hauling was properly started because of the buoyancy of a bag full of the fish. Today the catches are more mixed, with cod and coalfish being more prevalent than hake. Unlike otter trawlers which 'split' large bags of fish, pareja boats use a zipper line or removable lacing to open the bag when the catch is heavy. Fish are brailed out from this opening, as from a ring net, until a small quantity is left and this is taken aboard in the normal way.

Although pair fishing involves only two boats, three, five or more vessels may operate as a team so that the others are busy fishing while one is taking fish on board, or running fish back to the market. The system is not unlike that used by midwater pair trawlers in Scotland.

Distant Water Operations

The pareja boats working the cod grounds in the west Atlantic have discarded the traditional seine-like net. Instead they use a trawl similar to the conventional modern deep sea otter trawls. It is made of nylon and/or polyethylene and has a bigger mouth opening but much shorter wings than the traditional pareja. Long bridles and cables are still used, however, and the principle of sweeping a wide area still applies. For rougher grounds the nets may carry bobbins like the pair cod trawls used in the North Sea (Fig. 5).

These modern distant-water pair trawlers are equipped to freeze or salt their catch. They may also have liver oil extractors and drinking water evaporators. Trips may last as long as 60 days. Earlier salt-catch parejas stayed at sea for up to three months.

The character of Spanish pair fishing has changed a lot from the pre-war days when hake were abundant but it remains an efficient method of deep sea trawling which is very suitable for fleet operation

21

as are most methods of pair fishing. Similar gear is used by Portuguese and Italian vessels under the name of 'paranzella trawl'.

Pareja trawling was introduced to the United States California coast as early as 1876. It was used there for more than half a century but by the time of the Second World War, it had been replaced by otter trawling. The vessels used were small and low-powered and it is not surprising they were replaced. The Spanish fleet by contrast, was modernised over the years, and still fishes profitably off the east coast of America. As many as thirty large pareja trawlers have been observed at one time by USA aerial patrols. These vessels measure over 130 feet in length and tow trawls of up to 50-fathom headline length. The main area of operation off America is George's Bank which lies between Nova Scotia and New England, extending south as far as latitude 40°N.

Distant Water Vessels

By 1970 Spain had a fleet of over 200 'bacalao' or cod trawlers operating mostly in the west Atlantic. Over sixty per cent of these were parejas, using two-boat cod trawls. The most modern vessels freeze all of the catch, but salt processing is also still used. A modern pareja boat may measure from 30 to 50 metres in overall length, and be equipped with a diesel engine of from 500 to over 1,000 hp. Fish hold capacity ranges from 100 to 500 tons (180 to 700 cubic metres) (Fig. 6).

The largest and most modern pareja vessels resemble deep sea stern trawlers, and they are easily convertible to single-boat otter trawling. Cod are still salted on such vessels but this method is gradually being replaced with freezing.

In 1969 the shipyard Astilleros Construcciones completed two pairs of stern fishing parejas for Spanish companies. The smaller pair *Cristobal Colon* and *Aurelia Manuel* were 540 ton 1,250 hp vessels with an overall length of 170 feet. The larger pair *Bahia de San Sebastian* and *Bahia de Guipuzcoa* were 670-ton 1,500-hp vessels with a length of 180 feet overall. The ships were equipped with cod heading and splitting machines and could carry up to 425 and 600 tons of salt cod respectively. Both pairs carried crews of 29 men on each vessel

31·7 X 7·0 X 3·5 Metres   460 hp Heavy duty diesel engine

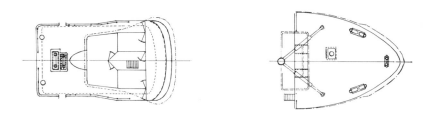

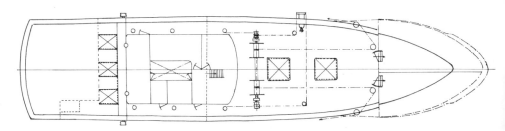

Fig 6   Modern multi-purpose Spanish vessel for pareja and seine fishing

and had a voyage endurance of up to 50 days without refuelling. A sketch of one of the 540-ton parejas is shown in Fig. 7.

Success Factors of Pareja Gear

Since the parejas were the first successful two-boat trawlers to be widely used, and as they have been operating for over 50 years, it might be appropriate to consider here some of the reasons for the success of pair trawling. Several contributing factors clearly emerge and it is interesting to note that in general they hold true for all methods of pair trawling regardless of the area of operation or fish species caught. Midwater pair trawling is somewhat different and cannot be compared directly with two-boat bottom trawling. The additional factors which contribute to its success are discussed in Chapter seven.

The elimination of otter boards contributes immensely to the towing efficiency of pair trawls. Tests and calculations by fishing gear technologists have indicated that otter boards create around 40% of the drag load on a single-boat trawler. Pair trawlers do not require otter doors and they consequently may pull a net which is nearly twice as large as that towed by a single vessel with an engine equivalent to the combined horse-power of the pair.

One natural effect of using a larger net with a

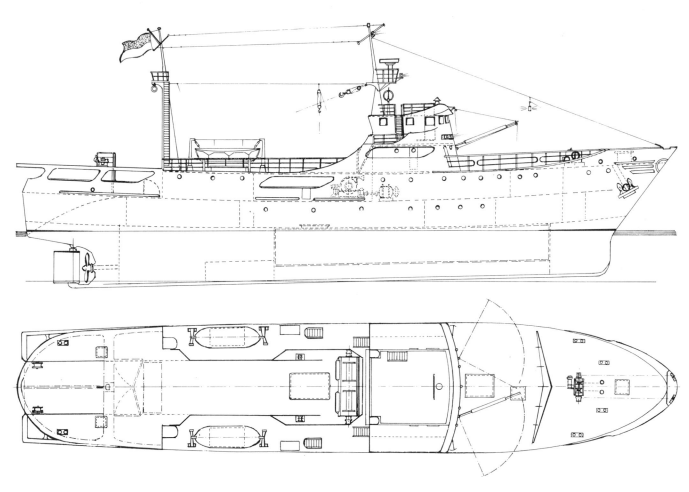

Fig 7   Large modern salt cod stern fishing pareja trawler

23

longer headline is that the vertical mouth opening, or headline height, can be much greater. This is especially useful for herring and for hake fishing, and also for cod at times when they may swim some distance above the sea bed.

Pair trawlers can sweep a much wider area of the seabed than can otter trawlers twice their size. The long warps on either side of the net guide or herd the fish in towards the net mouth. This herding effect, so familiar to Danish seiners, is often overlooked when discussing the pair trawl. Yet it is probably one of the most significant factors in its success. Successful pair trawling techniques in places as far apart as Spain, Denmark, Canada and Japan, include the sweeping of a wide area of the sea bottom. The one exception is when the fish are densely congregated on a patch of rough ground, as in the winter fishing for cod in the North Sea.

The fact that pair trawlers do not sail directly over the fish they catch must also have a positive effect on the fishing results. Perhaps the effect is minimised in deep water, but noise and vibration do tend to disturb fish, and in the case of an otter trawler, could alarm them in time to escape before they are aware of the otter boards or net.

Speed is not a critical factor in bottom pair trawling (though of course it is important in midwater fishing). In fact, when British steam trawlers first tried to fish for hake alongside Spanish parejas, they discovered that their catches declined rapidly if they towed too fast. Apparently a leisurely towing speed is best to obtain the gentle herding effect on the fish. This is also true for the bottom pair seining practised in Canadian waters.

Like any other modern technique, the application of pair fishing is not in itself a magic formula for fishing success. To succeed with any kind of fishing operation requires much hard work, skill and attention to the innumerable factors governing fish capture. Each method of pair fishing described in this book was developed to meet a specific need and to play a specific role. The delicate balance between the type of vessel, the operating conditions, the fish sought, and the market supplied must not be upset, or disastrous results could follow. This holds true for any type of fishing venture one wishes to pursue.

2 Bull Trawling

Japanese pair trawlers are known as 'bull trawlers'. They operate mostly west of longitude 135° E, in the Sea of Japan, the Yellow Sea and the China Sea. There are several hundred of these vessels in Japan and large fleets also operate from Korea and Thailand. Bull trawling appears to have been developed to enable smaller vessels with limited power to engage in deep sea trawling. The vessels themselves resemble Japanese 'Danish seiners' and the method of pair trawling usually involves sweeping an area of the sea bed as is done with the Danish seine. Fish catches are mixed, including a variety of bottom species, flats and round fish, shrimp, and some mackerel.

Bull trawl vessels are of traditional layout with engine room and bridge aft of amidships, forward of which the fish room is located (Fig. 8). They usually have a whaleback or raised deck forward and the net platform is situated aft. In all this they resemble the Spanish pareja boats. Small bull trawlers sometimes

pull the gear in over the stem as do some Japanese Danish seiners. But the winch arrangement is quite different to that on the Spanish boats. Most bull trawlers have the winch located directly above the main engine inside the casing or deck house. One drum protrudes from the casing on each side. The warps are hauled in over the stern roller and round the winch drums but then they are led from there on to warp reels on the fore deck from which they are later paid out, when setting the gear. In the early days of bull trawling in Japan and Korea it was not unknown for only one vessel of a pair to be equipped with a winch. The other boat was used only to assist in towing the gear.

The vessels vary in size from 60 to 90 feet in length with the bulk of them falling within the 70 to 85-foot bracket. Compared with otter trawlers they are rather lightly powered having mostly 250-350 hp engines. A crew of up to 12 men was carried on each vessel but these numbers have been reduced now with more mechanisation and automation. Trips last up to three weeks. Some of the vessels are dual purpose and may be converted to drift netting, Danish seining or otter trawling if required.

Dimensions of Bull Trawlers

Measurements of Japanese Bull Trawlers (dimensions in metres)

Type	L x B x D	Tonnage	Main engine hp
Modern, bridge forward	39.9 x 7.3 x 3.4	193.1	630
Modern, bridge forward	32.7 x 6.5 x 3.0	144.2	900
Modern, bridge forward	30.0 x 6.7 x 3.2	144.8	1,000
Traditional, steel	29.2 x 5.4 x 2.7	108.2	270
Traditional, steel	26.5 x 5.4 x 2.5	96.0	310
Traditional, steel	25.0 x 5.1 x 2.5	84.8	250
Traditional, wood	24.7 x 4.8 x 2.7	74.4	250
Traditional, wood	24.4 x 4.9 x 2.7	80.0	250
Traditional, wood	23.1 x 4.5 x 2.6	71.5	170

Source: *Fishing Boats of the World I; New Fishing Boats in Japan*, vol. 3

Plate 3   Small GRP bull trawler undergoing trials
Yamaha Motor Co

Some modern bull trawlers have been constructed on lines similar to stern trawlers. These vessels have the bridge forward and have a stern ramp and net drum on which the gear is taken in. They may have much more power than traditional bull trawling vessels and are equipped with refrigerated fish rooms. These ships can operate as single otter trawlers and can also engage in two-boat midwater trawling. Some large modern bull trawlers fish as far away from Japan as the Bering Sea.

The nets are almost all of the box-type design with long wings which are supported by many glass floats. A peculiar feature of Japanese bull trawls is that an extra float line is often fitted to the back of

Fig 8   Traditional Japanese bull trawler
Fishing Boats of the World 1

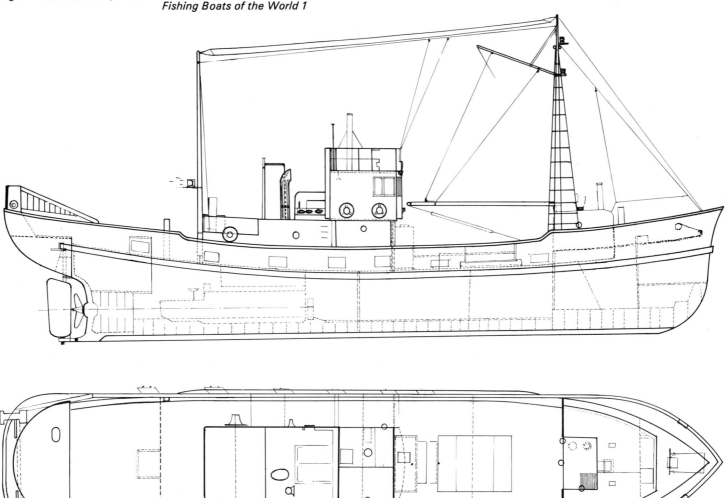

the square or overhang, and along the shoulder side seams. This is never done on European trawls. Mesh sizes are large in the wings where they may measure up to one foot. They decrease quickly at the shoulders and square where they may be only 2 to 3 inches. At the cod end, mesh sizes of around 1½ inches are common. The early trawls were made of manila but nylon (polyamide), polyester, polyvinyl alcohol and other synthetic twines are now used. There is considerable variation in the designs of the four-seam bull trawl and many experiments have been conducted in Japan to increase the mouth opening and reduce the towing resistance. As with most Japanese trawls there is a tendency to use many net panels of varying mesh size rather than a few tapered panels as on European gear. The former method involves more labour in construction and

complicates on-ship repairs for the crew members. Like the pareja boats, bull trawlers operate mostly on smooth grounds and only light chafing ground ropes are required.

The nets used on Korean vessels closely resemble the Japanese gear. In Thailand there is more variation, the tendency being to have relatively longer bags and shorter wings. Sometimes there is no overhang, headline and footropes being almost equal. Two-seam nets as used by otter trawlers are sometimes employed on Siamese pair fishers.

Operation

The operation of a bull trawl is fairly straightforward. The net is set out aft and is hauled in over the stern or side except on some smaller boats, where hauling is done over the stem. The

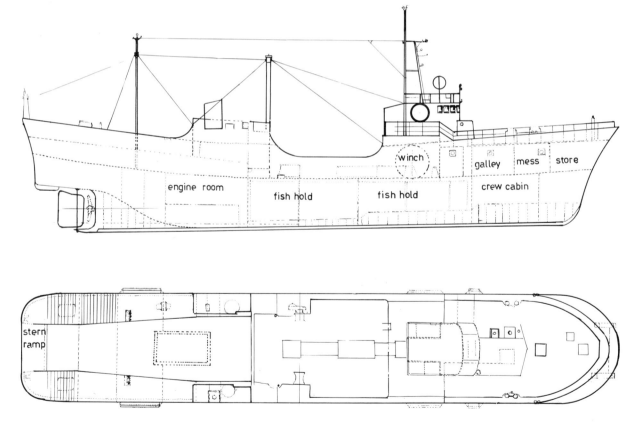

Fig 9   Modern Japanese bull trawler   *New Fishing Boats in Japan JFA*

method of shooting the warps is unusual. The vessels normally set the cable parts of the warp at right angles to the direction of tow. The remaining wire is paid out in the towing direction, thus forming a 'square' shape. Vessels may set the gear in a straight 'V' shape and some Japanese fishery scientists claim this is better for the shape of the net, but they ignore the sweeping action of the warps, which can herd fish in towards the net. Chains or weights may be added between the cable and warp to keep it on the bottom. The cables are made of thick manila or combination rope. At one time all of the warps were of such material, varying in thickness from the boat to the net, but wire warps are practically universal now. Warp length varies with the depth. At one time bull trawlers had difficulty operating in deep waters owing to the bulky nature of the cables and rope warps. Warp reels and wire warps now used make it possible to set in deeper waters. Total warp length of from five to seven times the depth is preferred but in

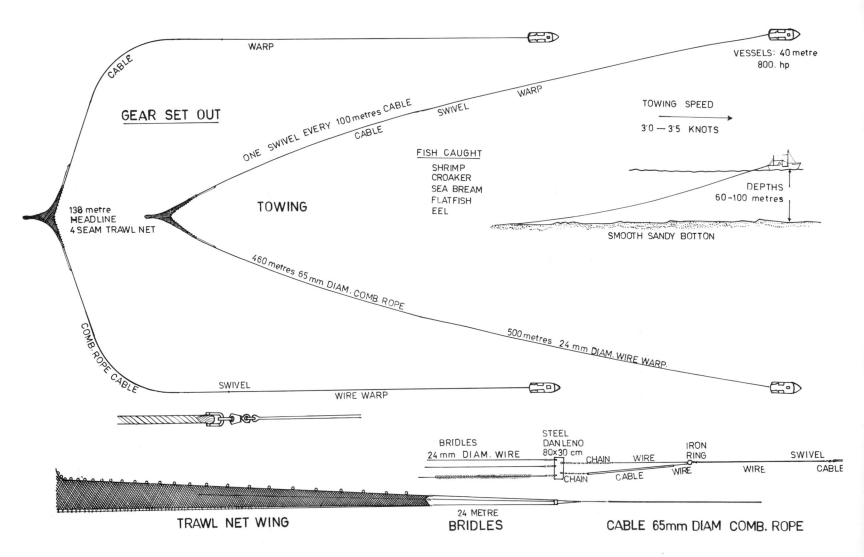

Fig 10 Operation of the bull trawl   *Tumilty WFA*

Plates 4a   Traditional bull trawler operations off Korea
and 4b   The warps come in over the stern roller aft
              but the cod-end is taken aboard forward
                                  Nam Yang Fishing Net Co

deeper parts a smaller ratio may be used (Fig. 10).

Towing speed varies from 1.5 to 3.0 knots or more, with 2.5 knots being a more common average. The gear is normally towed with the wind or current and tows last from 2 to 4 hours. The boats maintain a distance apart which is roughly half the total warp length. For instance, with 750 metres of warp paid out by each vessel, the boats remain 300 to 400 metres apart. Tests using transducers on bull trawls indicate that the distance between the boats is not critical. The net form and mouth opening is more affected by changes in the towing speed. As the speed increases the headline height is reduced. A typical bull trawl with a 230-foot headline, pulled by two 350 hp vessels may have an average headline height of around 10 feet. This can rise to nearly 20 feet at speeds of 1½ knots which is too slow for commercial trawling. As towing speed increases towards 3 knots the headline may come as low as 2 feet. The net is therefore not suitable for high speed trawling. This may not be unusual for a bottom trawl as, with the exception of herring trawls, high towing speeds may be disadvantageous. This is particularly true of sweeping-type gear such as the bull trawl, Danish seine or Canadian pair seine. With such gear a skipper may often improve his catches by towing slower rather than faster.

Bull trawls are generally heavier than pareja nets. Both these types carry many glass floats and have manila-wrapped groundropes with chains attached at intervals. The Japanese nets also have a similar dan leno or bridle pole. Bull trawl cables are of manila rope up to 5½ inches in circumference. Such thick warps are sometimes also used on Japanese Danish seines; 150 fathoms of the cables may be used on either side of the net. Wire warp sizes normal for the tonnage and power of the trawler are from ⅝ to ¾ inch diameter.

A notable feature about the operation of bull trawlers is the peculiar master ship and assistant vessel arrangement of the pairs of trawlers. This kind of system is also used in some two-boat purse seine fishers in the Orient. Each pair of vessels consists of one master ship and one slave vessel. The vessels will be of the same size and power, but the master ship carries more electronic equipment than

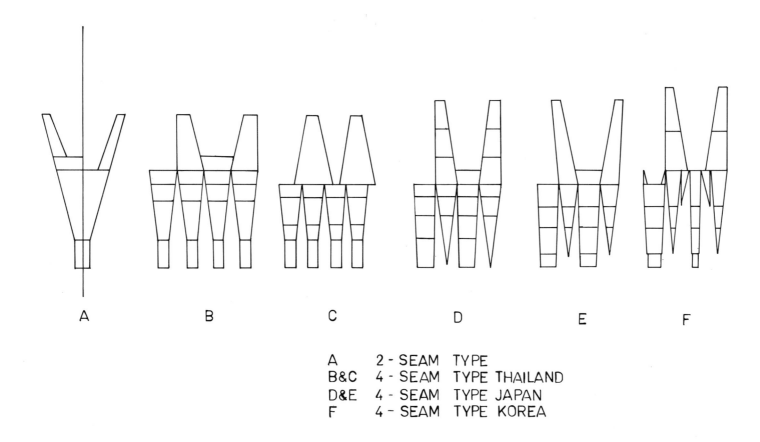

A   2 - SEAM   TYPE
B&C   4 - SEAM   TYPE THAILAND
D&E   4 - SEAM   TYPE JAPAN
F   4 - SEAM   TYPE KOREA

Fig 11   Bull trawl design types

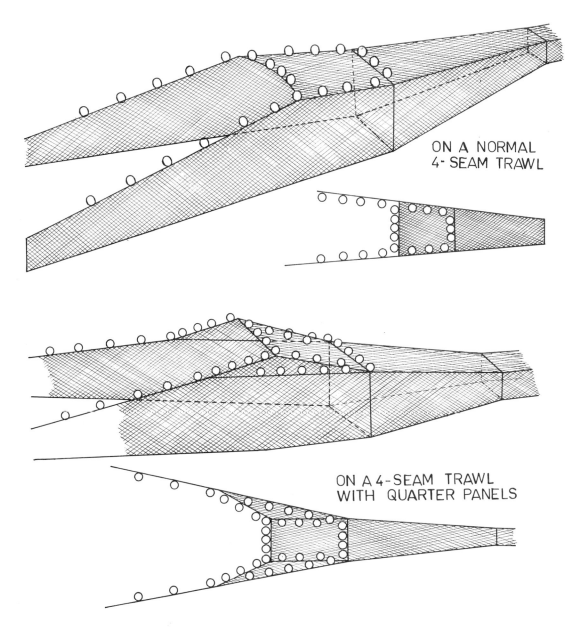

ON A NORMAL
4-SEAM TRAWL

ON A 4-SEAM TRAWL
WITH QUARTER PANELS

Fig 12   Illustrating how Japanese bull trawls may be rigged with an extra float line around the square

31

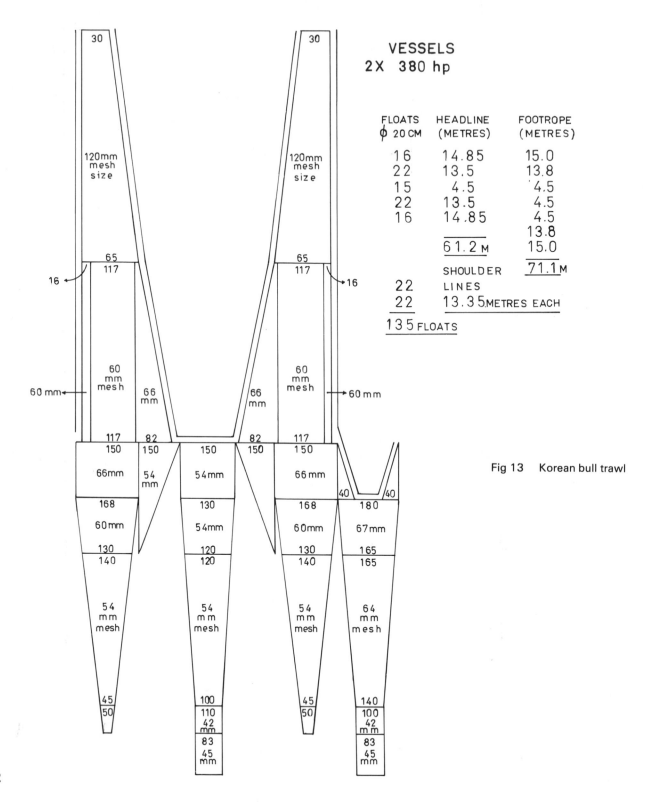

VESSELS
2X 380 hp

FLOATS φ 20 CM	HEADLINE (METRES)	FOOTROPE (METRES)
16	14.85	15.0
22	13.5	13.8
15	4.5	4.5
22	13.5	4.5
16	14.85	4.5
		13.8
	61.2 M	15.0
	SHOULDER	71.1 M
22	LINES	
22	13.35 METRES EACH	
135 FLOATS		

Fig 13   Korean bull trawl

32

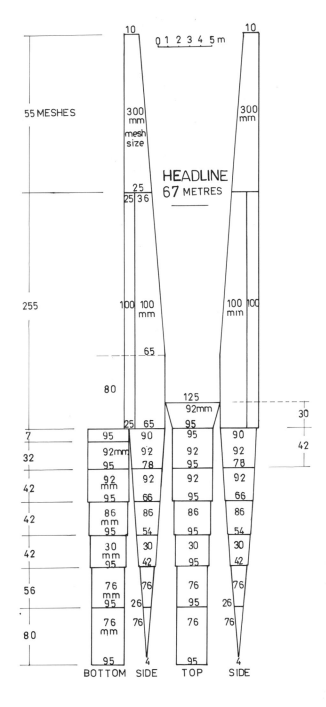

Fig 14   Japanese bull trawl   *F Bourgeois (le chalut)*

its assistant vessel. In former times when bull trawlers were smaller and less well-equipped, this meant that only one of the boats would have a radio and an echo-sounder. Nowadays, all of the vessels in a modern fleet would have this basic equipment, but the master vessel has more than its partner. It may have all the position-fixing equipment such as Loran, RDF and radar. While the partner vessel may only carry a small vhf radio-telephone, the master ship will have a powerful long-distance radio telephone in addition to its vhf set. The master ship carries a senior skipper or fishing master who directs the operations of the pair of trawlers. He is responsible for checking the position and for making decisions relative to the fishing tactics. While liaising with the fishing masters of other pairs in the fleet, he will decide where to shoot, in which direction to tow and what modifications to make to the gear. On large vessels, a radio operator will be carried in addition to a fishing master.

The master/slave partnership arrangement is not so common on European pair fishers, but this is probably because most of them are privately owned, while the large fleets of bull trawlers are mostly company owned. European fishermen prefer the single skipper arrangement with one man responsible for navigation and fishing activities, but until recently, many Japanese vessels carried two skippers, one a navigating captain, and the other a fishing captain. The bull trawl pair-fishing system probably suits their type of bottom trawling in which fish detection does not play a large part and pairs stick together throughout each trip. Since many dozen pairs of trawlers must operate in concert, it is better to have only one skipper in each pair responsible for making the decisions.

Larger Bull Trawlers

Large modern bull trawlers of the stern fishing type use heavier gear and a much larger net than the traditional vessels. Ships in the size range of 120 feet, 200 tons, with 600–1,000 hp may use a trawl net with a headline measuring 450 feet or more. They are probably the largest trawl nets in use anywhere in the world. The mouth opening is much higher than the 20 feet achieved by traditional bull trawls, and

33

this helps in the capture of high swimming *Penaeus* shrimp and sea bream. Towing such a large net, the vessels may be 700 metres apart when travelling with the tide, but they prefer to tow only 500 metres apart if towing into the current. The ground cables used with this gear are extremely heavy. Up to 460 metres of 8-inch circumference combination rope may be used on each side of the net. Inside this rope is a core of 1-inch diameter steel wire rope. The warps are also made of this size wire. These large vessels tow at 3 or 3½ knots depending on whether they are working primarily for shrimp or fish. One surprising aspect of the operation of the gear is the speed with which it can be set and hauled. Shooting may take only ten minutes, and hauling, just over half an hour. Vessels in a pair alternate, each setting its net in turn. The vessel setting the net will haul both warps and cables. This means that after hauling, one boat will have both its warp and cable reels full while the other will have one of its reels empty.

Similarities to Pareja Fishing

Bull trawling in the Orient resembled pareja trawling more than any other kind of two boat fishing – particularly as it was worked by medium-sized vessels in the 1950s and early 1960s. Both types of gear were towed at moderate speeds over smooth grounds, by vessels of up to 160 tons, 500 hp. The Japanese nets were rigged to take a greater variety of species however. In recent years, bull trawling has developed into a more sophisticated deep sea fishery involving large fleets of powerful modern trawlers using expensive heavy duty gear.

There are fleets of bull trawlers operating in the China and Japan Seas from south of Thailand to north of Japan and south-east as far as West Irian – New Guinea. The fishery probably reached its peak in terms of numbers of vessels in the late 1950s but it still accounts for a considerable amount of fish. In recent years there has been a trend toward otter trawling, but wherever there are large stretches of smooth trawlable ground, the two-boat system appears to be preferred.

Bull trawlers operate in large fleets and may employ transport vessels to take the catches to port. A fleet of 80 vessels (40 pairs) may be served by four

such carrier vessels. The carrier boats do not interfere with the fishing operation, but come alongside the catcher boats while they are towing. Large inflatable fenders are inserted between the vessels before they are lashed together and the catches are transferred. The transport vessel system enables bull trawlers to remain at sea each for up to 40 days. Carrier vessels bring out food, provisions and mail to the ships. The fish are thus brought fresh to market, boxed, with ice, every five or six days. This aspect of bull trawl operations also resembles the activities of pareja trawlers in former years.

Fig 15   Small Japanese GRP bull trawler

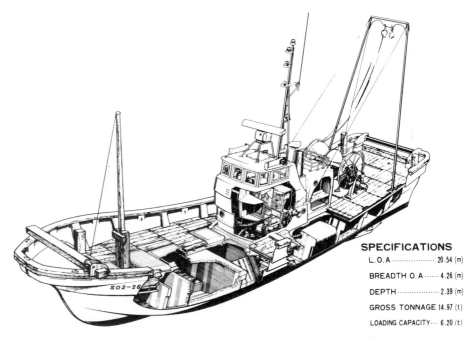

SPECIFICATIONS

L.O.A .... 20.54 (m)

BREADTH O.A .... 4.26 (m)

DEPTH .... 2.39 (m)

GROSS TONNAGE 14.97 (t)

LOADING CAPACITY... 6.20 (t)

Fishing Report:
Bull Trawling in the Yellow Sea

The pair trawler *Tokai Maru No. 27* and her partner vessel set sail from Nagasaki on their 40-day trip to the fishing grounds in the north Yellow Sea between South Korea and the mainland of China. It took only a day and a half to travel the 370 miles to the fishing grounds to join the rest of the company fleet operating in that area. During the voyage, the 15-man crew on each of the 125-foot, 200-ton vessels, prepared the gear and equipment for the fishing operations.

Once on the fishing grounds, the captains studied the sea bed lying 40 fathoms beneath. The bottom was chiefly composed of smooth sand, ideal for pair trawling. By means of their high frequency echo-sounders and fish scopes (crt units), the skippers also searched for traces of shrimp and fish marks. The surface and bottom sea-water temperatures were checked by means of a thermometer probe. For shrimp fishing in this area, temperatures around 10 or 11° Celsius (50-52°F) were considered best.

The vessels came alongside in preparation to shoot the gear, and the cod-end was dropped down the stern ramp. A slip hook was attached to the gear which was then pulled aft till the body of the net was in the water. The rest of the gear slid down the ramp until the huge 456-foot headline trawl net stretched out behind the stern. As soon as the 13-fathom bridles and iron spreader were in the water, the heavy combination rope warp was released. The partner vessel meantime had passed the end of her rope in readiness. Both vessels then sped away at full speed, paying out the heavy rope from their storage reels. When the 250 fathoms of rope were almost spent, both vessels turned on to parallel courses. The last of the rope then ran out followed by the 250 fathoms wire warp. The ships slowed down as the last of the wire ran out. Towing then started, at a speed of 3 knots, with the vessels over 600 yards apart, with the warp attached to a slip hook on the rail aft. With over 600 hp each, the ships had no difficulty in pulling the huge net and 500 fathoms of warp, either into or with the tide.

Once the captain was satisfied the gear was towing properly and the vessel on course, he handed over to the mate and went below for some food and a rest. The two ships remained in contact with each other by vhf radio.

After two hours the captain took command again, and within 30 minutes the vessels were drawing together in preparation for hauling. The crew appeared on deck at this point, complete with hard hats, life-jackets and protective clothing. A messenger wire was passed over and the end of the partner ship's warp was shackled into the *Tokai Maru* spare drum or storage reel. Hauling then started with the warps coming in over rollers on the gantry, around the whipping drums and back on to the storage reels. Within 12 minutes the spreaders were up. A few more hauls on the winch using independent wires, and the net and bridles had been hauled up the ramp on to the deck. The cod-end was still in the water though, and it was then hoisted up the ramp by means of a gilson wire. A sharp tug on the cod-end line and the fish were released into pounds on the port side. Out poured the shrimp and fish – croaker, sea-bream, yellow-tail, sharp-toothed eel, blow fish and other species.

While some crew men began to sort the catch, the others returned to attend to the gear, for the partner vessel had already put its trawl in the water. The end of the warp from *Tokai Maru* was passed and shot. Within minutes towing began once more and the fish handling started in earnest. After being sorted into species, the fish and shrimp were washed then graded on deck, the shrimps being graded into five different size groups. The graded shrimps were passed into the 6,000 cubic foot fish room which contained 40 tons of flake ice and hundreds of empty boxes. The boxes, holding 55 and 77 pounds of fish respectively, were lined with polythene and packed with alternate layers of shrimp and ice, and covered with polythene. The fish were packed bellies uppermost in boxes which were lowered into the hold where each was iced and covered with a sheet of greaseproof paper.

Once the fish were all packed, and the decks washed down, the crew relaxed briefly before hauling time. Fishing continued like this, day and night, almost non-stop, for six weeks. During night time the tows were one hour longer and the crew were able to get some sleep.

While towing on the sixth day, one of the company fish transport vessels came alongside. Huge rubber fenders were put over and the ships made fast together.

The boxes of fish and shrimp, over 1,600 of them, were transferred by a simple union purchase rig. As chilling pipes had kept the fish room temperature at 3°C (37.5°F), the fish were in excellent condition. The weekly visits of the transport vessel were looked forward to as it also brought stores for the vessel and mail for the crew.

Fishing continued throughout the trip with the company fleet of 40 pairs of trawlers operating within a ten mile radius. Occasionally the whole fleet would move farther afield when scouting vessels located better catches on nearby grounds.

After six weeks constant fishing, ice, fuel and stores were getting very low and the crew were longing for home. Finally the last tow was completed and the ships turned to steam to port together. The total catch of shrimp alone for each vessel, for the trip, amounted to over 4,000 boxes. This, plus the fish, assured the crews of a substantial pay packet on their return. With only four days in which to enjoy home comforts before the next voyage, they well deserved it. Within 36 hours the pair of bull trawlers had sailed through the east China Sea to dock in their home port in Kyushu Island, Japan.

(Adapted from the voyage report by J. Tumilty.)

The following illustrations show in broad sequence the fishing operations of a modern Japanese stern bull trawler
J Tumilty WFA Industrial Develop. Unit

Plate 5a   The trawl net on deck aft

Plate 5b   The steel spreader (dan leno)

Plate 5c   The warp storage reel

Plate 5d   The view aft when towing

Plate 5e   The warp held by towing hook aft

Plate 5f   Hauling in the warps

Plate 5h   The trawl net wings coming aboard

Plate 5g   Hauling in the heavy combination
                  ropes

Plate 5i   The cod end being hauled up the
                  ramp

3 Baltic Pair Trawling

In 1927 when Spanish pareja boats were beginning to fish off the west of Ireland, German and Polish cutters were successfully developing a bottom pair trawl of their own. It was developed for the Baltic Sea fishery, and enabled the vessels there to work year-round for cod and lucio-perch. Later the gear was adapted for herring fishing and it also proved successful in tropical waters off West Africa.

The Baltic Sea cutters were lightly-powered diesel-engined vessels suited for drift netting and some light trawling. They were really not powerful enough for otter trawling in the North Sea. The Baltic Sea is shallow, most of it being less than 40 fathoms in depth. It also has a much lower salinity than the North Sea, and this naturally results in some fish species not usually found in the Atlantic.

Pair trawling in the Baltic proved to be a more economical method of fishing for the cutter vessels. It was more than twice as productive as otter trawling and this amply compensated for the use of two vessels. The gear was worked in a similar way to the pareja trawl except that the boats shot and hauled the net as on a side trawler. Long ground cables and bridles were used and weights were attached at the warp-ends and the lower wing ends of the net. The nets were lightly rigged and they achieved much greater headline heights than otter trawls of a similar size. To maintain the good mouth opening, three bridles were used between the dan leno and the net. The centre bridle was attached to the mid-seam or selvedge as on a Scandinavian type wing trawl. Later the nets were constructed with the 'V' wing end more typical of high-opening wing trawls.

The German name for this type of two-boat bottom trawl is 'tuckzeese'. Though not as well-known as the pareja gear, its use has spread just as far afield. As fish catches declined in the Baltic Sea, the vessels began to operate in the North Sea where the gear was found to be equally effective, particularly for herring.

Adapted for Herring

Special tuckzeese trawls had been developed to catch Baltic herring when they were on the grounds, and this kind of net was adapted for North Sea conditions. The two-boat bottom trawling for herring in the North Sea continued from 1950 to 1970 but with declining catches and the development of midwater trawling the operations declined significantly. In 1967, however, under a bilateral aid programme, German tuckzeese trawls were tested off the Ivory Coast in West Africa. The two-boat trawls produced four to ten times as much fish as single vessel otter trawls in the area. They also caught a much higher percentage of large high-swimming fish due to their greater headline height. As a result of these experiments the method of two-boat bottom trawling looks very promising for certain tropical and semi-tropical fisheries.

The tuckzeese trawl is still used by small cutters in the Baltic with little variation in the gear from the 1950s except that synthetic twines have replaced cotton and a more modern net design has been evolved. The boats set and haul the gear like side trawlers. Most midwater pair trawlers operate this way except that they use two warps from each vessel whereas the bottom pair trawl needs only one warp per vessel. The net is put over the side of the cutter till it is all in the water. Some Baltic cutters are rigged for port side fishing so the gear may be set with that side to windward. As the bridles are being paid out, the partner boat comes alongside and receives the end of the combination rope cable which it connects to its warp end. The warp weight is also attached here. The remainder of the cables are put overboard and the vessels take the strain of the gear. Once the net is spread clear behind, the winch brakes

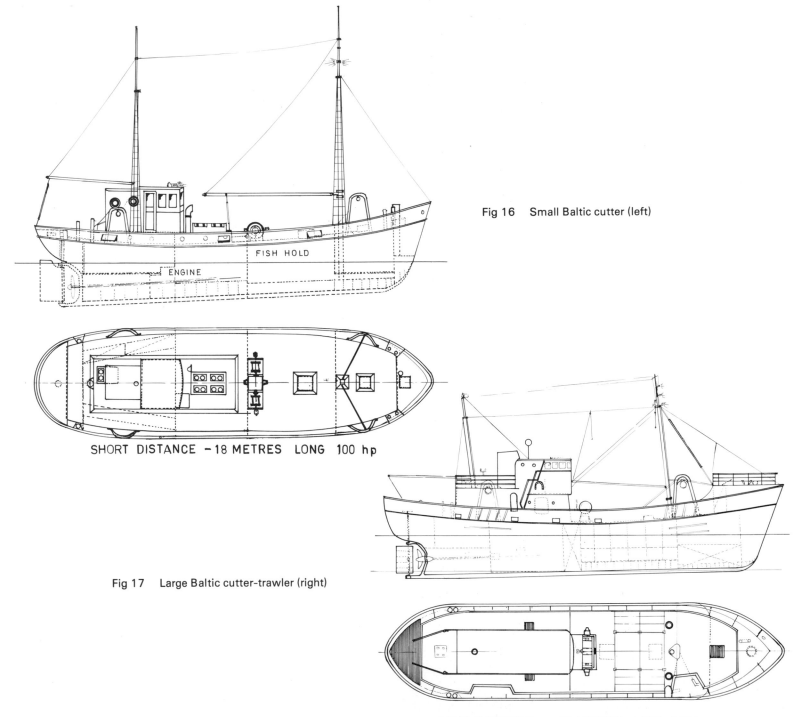

Fig 16   Small Baltic cutter (left)

FISH HOLD

ENGINE

SHORT DISTANCE – 18 METRES LONG 100 hp

Fig 17   Large Baltic cutter-trawler (right)

LONG DISTANCE – 27 METRES LONG 300 hp

are released and the boats shoot the warp on diverse courses. From 10 to 15 times the depth of wire is paid out in shallow water, and rather less in the deep. This is a similar practice to single-boat light trawling in which vessels rarely set less than 100 fathoms of warp in shallow water.

The Baltic pair trawl is a 'smooth-ground' gear like the pareja and unlike the cod pair trawl which is primarily used on rough grounds.

The ground rope consists only of a length of light chain attached at intervals to the foot. Sometimes iron rings are suspended from the footrope instead of chain. The gear is therefore very light and it skims easily over the bottom with the minimum of friction. The long sweeps or cables help to drive fish in towards the net. Part of the success of the gear in the Baltic may be due to the long distance between the vessels and the net and the fact that the boats do not pass directly over the path of the net as the engine noise could disturb the fish in shallow waters. Tows may last for 2 to 3 hours or more, but they are not as long as with pareja gear. The vessels haul in the normal way and after the warps and weights are up to the gallows, the bridle end is passed to the hauling boat which takes the gear in over the side.

The Baltic herring trawl which is also used in the

Fig 18   Baltic two-boat trawls

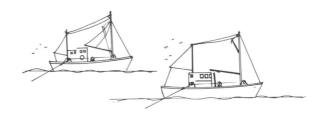

VESSELS, 2 X 20 METRE, 200 hp CUTTERS

WARP/DEPTH RATIO 10:1

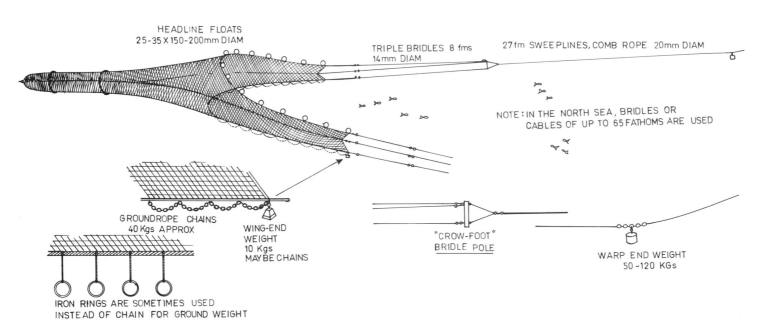

HEADLINE FLOATS
25-35 X 150-200mm DIAM

TRIPLE BRIDLES 8 fms
14mm DIAM

27 fm SWEEPLINES, COMB ROPE 20mm DIAM

NOTE: IN THE NORTH SEA, BRIDLES OR
CABLES OF UP TO 65 FATHOMS ARE USED

GROUNDROPE CHAINS
40 Kgs APPROX

WING-END
WEIGHT
10 Kgs
MAY BE CHAINS

"CROW-FOOT"
BRIDLE POLE

WARP END WEIGHT
50 -120 KGs

IRON RINGS ARE SOMETIMES USED
INSTEAD OF CHAIN FOR GROUND WEIGHT

41

North Sea resembles the cod and perch nets except that the bag and cod-end meshes are smaller and the body of the net is proportionately longer. Fishermen in Poland, Germany, Sweden and Denmark are well experienced in herring trawling and they were the first to develop the semi-pelagic type herring trawls. These are bottom trawls which are designed to skim the bottom rather than drag over it like a Granton trawl. They are also high opening nets and are thus able to catch fish swimming several metres above the sea bed. It is sometimes amazing how these nets will catch bagfulls of pure herring in areas where ordinary trawls take only demersal fish. Yet they are bottom nets and must never be confused with midwater trawls. The secret of their ability to catch herring lies in the rigging and hanging of the nets to obtain the optimum mouth opening and the minimum of towing resistance which would create a kind of slipstream or wave in front of the net. The bridles must be measured carefully each trip, and need to be adjusted regularly to ensure good performance. The netting panels on the top section must also be kept tight relative to the lower section.

Plate 6a   A 20-metre wooden trawler   *Iver Christensens*

Daylight Fishing

Herring bottom trawling is usually conducted in the daytime on clean bottom in depths up to 100 fathoms. At night the fish rise towards the surface and can only be caught by midwater trawl or purse seine, but in the daytime they can be found near the bottom. Unlike their nocturnal behaviour when the herring schools are concentrated, in daytime the fish are scattered and may be detected on the echo-sounder only as a faint haze or in 'smoky' marks near the bottom. The tows are similar in length to those by otter trawlers, usually 2½ to 3½ hours.

When working the deeper waters with the herring trawl the cutters may use 65 fathoms of ground cable or sweeps, plus bridles. Warp length would vary from five to ten times the depth depending more upon the amount of warp the winches held than anything else. The cutter vessels have only 150-220 hp each and for herring trawling a considerable amount of power is normally required. Nevertheless, they are able to pull a net with 210-

Plate 6b   A 25-metre steel trawler   *Iver Christensens*

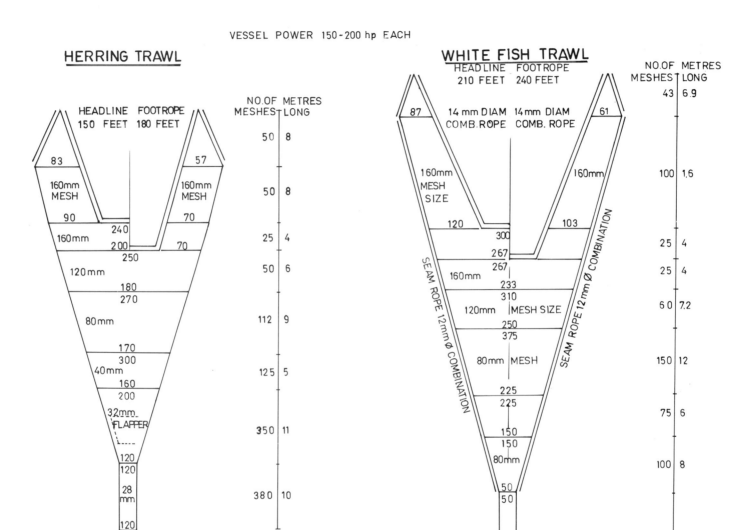

Fig 19   Bottom pair trawls for 150 hp/200 hp cutters

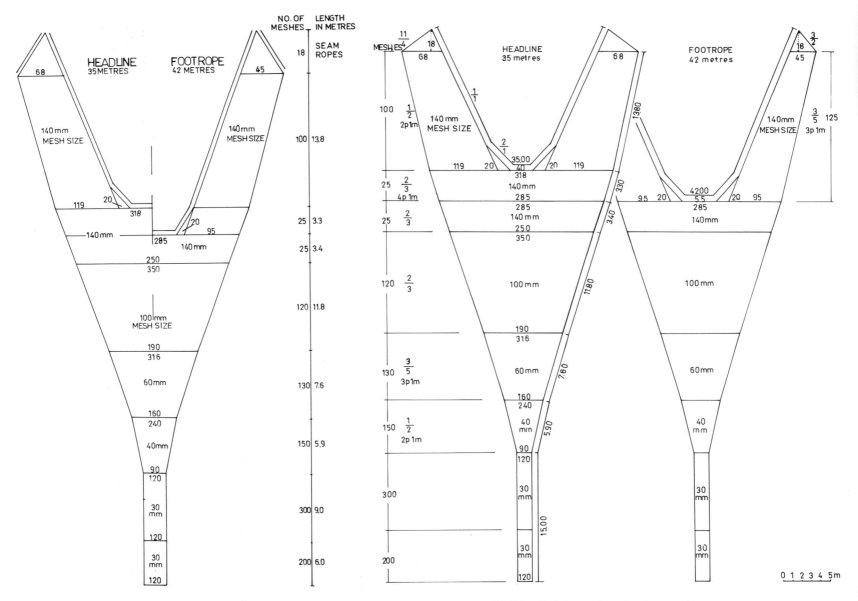

Fig 20 Early Baltic herring bottom pair trawl

Fig 21 Polish two-boat herring trawl
Klima, Bruski, Netzel (le chalut)

44

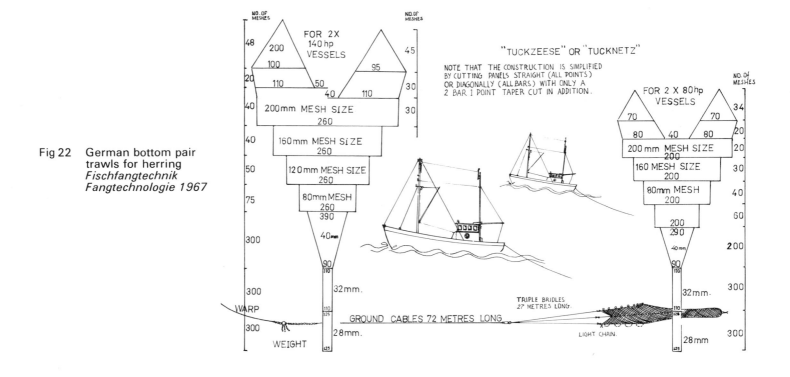

Fig 22   German bottom pair
          trawls for herring
          Fischfangtechnik
          Fangtechnologie 1967

FOR 2X 140 hp VESSELS

"TUCKZEESE" OR "TUCKNETZ"

NOTE THAT THE CONSTRUCTION IS SIMPLIFIED BY CUTTING PANELS STRAIGHT (ALL POINTS) OR DIAGONALLY (ALL BARS) WITH ONLY A 2 BAR 1 POINT TAPER CUT IN ADDITION.

200mm MESH SIZE
160mm MESH SIZE
120mm MESH SIZE
80mm MESH
40mm
32mm.
28mm.

WARP
WEIGHT
GROUND CABLES 72 METRES LONG
TRIPLE BRIDLES 27 METRES LONG.
LIGHT CHAIN.

FOR 2 X 80hp VESSELS
200mm MESH SIZE
160 MESH SIZE
80mm MESH
40mm
32mm.
28mm

Fig 23   Modern Baltic pair
          trawl

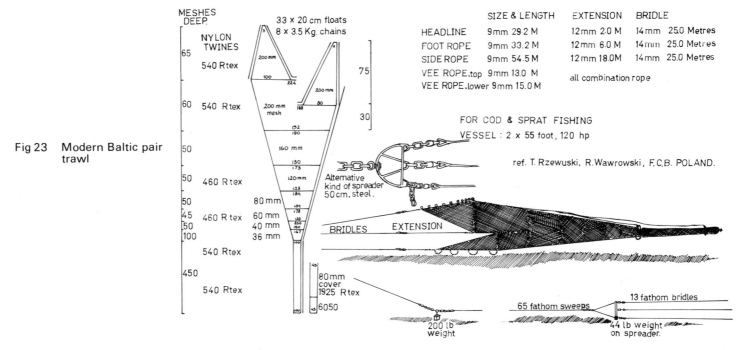

MESHES DEEP.
NYLON TWINES
540 Rtex
540 Rtex
460 Rtex
460 Rtex
540 Rtex
540 Rtex

33 x 20 cm floats
8 x 3.5 Kg. chains

200mm mesh
160 mm
120 mm
80mm
60 mm
40 mm
36 mm

80mm cover 1925 Rtex

	SIZE & LENGTH	EXTENSION	BRIDLE
HEADLINE	9mm 29.2 M	12mm 2.0 M	14mm 25.0 Metres
FOOT ROPE	9mm 33.2 M	12mm 6.0 M	14mm 25.0 Metres
SIDE ROPE	9mm 54.5 M	12mm 18.0M	14mm 25.0 Metres
VEE ROPE,top	9mm 13.0 M		
VEE ROPE,lower	9mm 15.0 M	all combination rope	

FOR COD & SPRAT FISHING
VESSEL : 2 x 55 foot, 120 hp

ref. T. Rzewuski, R.Wawrowski, F.C.B. POLAND.

Alternative kind of spreader 50 cm, steel.

BRIDLES   EXTENSION

200 lb weight

65 fathom sweeps   13 fathom bridles
44 lb weight on spreader.

foot headline and a mouth opening that may be over 20 feet high. The headline height is greater than that achieved by the bottom herring trawls operated by 150-foot stern trawlers which use 'kites' to scare fish down into the net. No doubt the cutters are able to tow such a large gear because they do not have to cope with the drag of otter boards which may amount to nearly half the load on an otter trawler. Also they have an extremely light gear that creates very little friction on the sea-bed.

Use in the Tropics

The success of the Baltic trawl in tropical waters is of particular interest to developing countries where the fishing vessels may be small and lightly powered. It may enable little boats to engage in bottom trawling formerly prosecuted by much larger vessels only. Not only that, but the high opening Baltic net may take larger quantities of high swimming fish than the otter trawl. Thus the gear can be both economical and productive, requiring lower capital investment and operating costs, yet resulting in higher catches.

Pair trawling in the Baltic at present times appears to concentrate on sprat and cod fishing. Some very high opening cod trawls are now in use. They differ from the Danish two-boat cod trawls in that they are essentially smooth ground nets while the Danish North Sea trawls are designed primarily for rough grounds. A modern Baltic pair trawl for cod and sprat may measure 400 x 8-inch meshes around the mouth, and have a headline length of 100 feet. The high opening nets may be of three-bridle design, each bridle being about 15 fathoms long. While the bridles may be of combination rope, the cables may be made of nylon rope, one inch diameter. These cables may be 60 or 70 fathoms long. A 90-kilogram weight is attached between the cable and the warp. The whole gear is towed by two 120 hp vessels. This kind of net is so similar to the Danish cod pair trawls, apart from the groundrope gear, that it might well be considered to be its predecessor.

Plate 7   A Grimsby cod pair trawler the Skanderborg: a traditional wooden vessel   *T J M Wood*

4 Pair Trawling for Cod

With the decline in pareja fishing and bull trawling one might have thought that two-boat bottom trawling was soon to be a thing of the past. However, in the late 1960s Danish fishermen put their heads together and developed a method of pair fishing that has been astounding in its success.

The Danes were past masters at fishing the North Sea in small boats. Since 1920 their little wooden seine netters had been anchor-dragging for plaice and cod from the Dogger to the Bergen bank. They knew every inch of the North Sea grounds and had a wealth of experience in following the seasonal movements of fish stocks. Plaice catches had declined considerably since the war, and cod and haddock were becoming predominant in the seine net catch. Cod are a unique fish with seasonal migrations and spawning behaviour unlike haddock or hake. Spawning takes place during the winter in the North Sea and it is then that the mature cod congregate in dense shoals. In the early part of the season these schools tend to be predominantly male or predominantly female, then they become more mixed as spawning takes place. Cod prefer sandy bottom with shells or shingle, where they can feed on sand eels, small fish and some crustacea. They also prefer to spawn on such grounds. But like most bottom fish they feel insecure on clean bottom and for protection they stay most of the time on hard, rocky grounds. This of course makes them very difficult to catch in a bottom trawl. The Danes had often seen these large schools of cod on their echo-sounders and they began to devise a way to net them.

It is possible for deep sea otter trawls to come over rough grounds with a minimum of damage to the net, but these are heavy Granton trawls with iron bobbins, and they are towed by vessels 140-240 feet in length with 1,000-2,000 engine hp. The Danish seiners were only around 65 feet long with 114-230 hp engines, and they could not hope to pull such heavy gear. However, the potential rewards were so great, they persevered and developed a suitable trawl.

First they reckoned that a large wing trawl would make a suitable net, something resembling their seine net wing trawls, but much bigger. The meshes in the wings and mouth would be large, to give low towing resistance and allow a big net mouth circumference. To keep the net off the rough bottom was a major problem. They would need extra large bobbins, 16 or even 20-inch diameter but only the largest of deep sea trawlers could handle them. So instead of iron bobbins, they elected to use plastic. At first extra large floats were simply modified to make bobbins but later, proper heavy duty plastic bobbins were produced. They had holes bored in them to permit them to fill with water so they would sink. But on deck they were light and easy to handle. Once the net and ground rig were designed it was left for the skippers to develop the best fishing techniques.

Unlike the large deep sea trawlers, the Danish boats preferred to use a large warp to depth ratio. It could be as high as 10 : 1. In addition they fitted long ground cables ahead of the net and added a heavy weight between the cable and the warp. The net result of all this is a semi-seining type operation with fish being herded towards the net from either side. It also put less sudden strain on the light gear when working in heavy weather. The net carried a lot of floats and as the plastic bobbins were of almost neutral buoyancy, the gear touched just lightly on the bottom. This is exactly what the skippers wanted. There was little bottom resistance and only a limited chance of snagging on the ground. The large meshes made the net easy to tow and the combined power of the little boats proved sufficient.

What happened then is now fishing 'history'. In Grimsby, England, these 60-foot boats came in after short trips, laden down with catches of cod that many a vessel twice the size would have been glad to produce. As the skippers gained experience with the gear, their catches grew even larger, and their trips shorter. By the end of 1973 some anchor-seiners had doubled their previous annual grossings – a leap from £80,000 to £160,000. In one eight-day trip that year, a pair of vessels landed over 900 kits of fish to gross over £10,000 each.

Later, from a four-day trip they grossed over £5,000 each. These kind of earnings compare favourably with the daily averages of 180–240-foot deep sea trawlers operating around Iceland, and it completely disproves the argument that pair fishing is uneconomical because two vessels must share the proceeds from one net. If one compares the annual proceeds from a 'one net' pair cod trawl with the annual proceeds from a 'one net' deep sea Granton trawl it will be clearly evident which is the most productive, and the pair trawl requires only 2 x 40 ton 150 hp vessels to pull it. The deep sea trawl needs a single 400-ton 1,000 hp vessel. For these and other reasons, many large British trawling companies decided in 1973 to invest heavily in new fleets of seiner-pair trawler vessels.

Thus was born pair trawling for cod, a unique method of 'aimed trawling' which enables small vessels to work rough grounds and use a net of a size

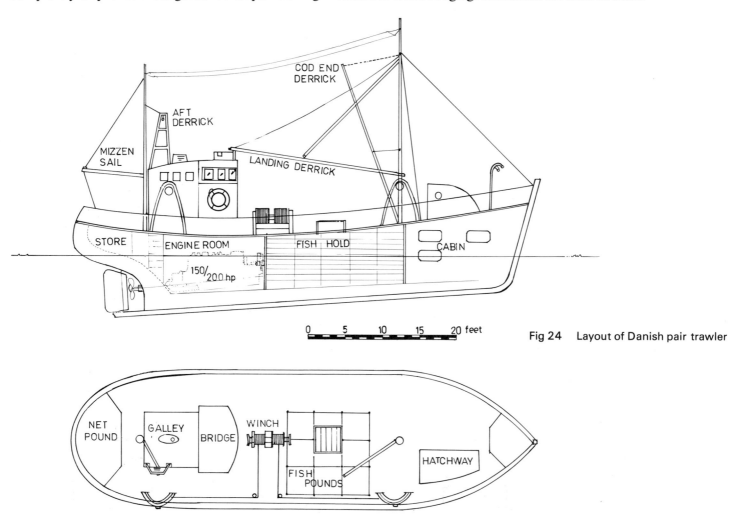

Fig 24   Layout of Danish pair trawler

equivalent to that on the largest deep sea stern trawlers. As the cod fishing is a seasonal one, the boats normally revert to anchor-seining for the remainder of the year. Some of them now also engage in midwater pair trawling for sprat and herring when these fish are around. The more successful pairs now operate throughout the year.

Vessels

A typical cod pair trawler measures 65 x 18 x 9 feet and has a main engine of about 170-230 hp. The engine room is aft, fish hold amidships, and crew accommodation forward. This is a traditional Danish arrangement used on their anchor seiners and is still preferred on the pair trawlers. The galley may be situated in the cabin or in a small mess room behind the bridge. The wheelhouse is situated well aft and modern vessels have a whaleback on the foredeck. In accordance with Danish tradition, the winch lies fore-and-aft on the deck though some anchor-seiners have their trawl winches athwartships. Power blocks are now used for hauling in the net. They may be single sheave davit type, or multiple sheave with a fixed pedestal. The centrally located fish room enables these small boats to carry up to 40 tons of fish. The net is carried aft or on the starboard side deck and the fish are taken aboard amidships. Pair trawlers and anchor-seiners

Fig 25   Cod pair trawl operation

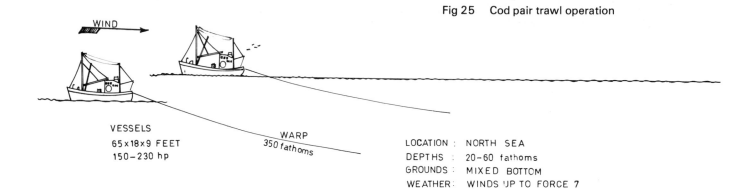

VESSELS
65 x 18 x 9 FEET
150-230 hp

WARP
350 fathoms

LOCATION : NORTH SEA
DEPTHS : 20-60 fathoms
GROUNDS : MIXED BOTTOM
WEATHER: WINDS UP TO FORCE 7

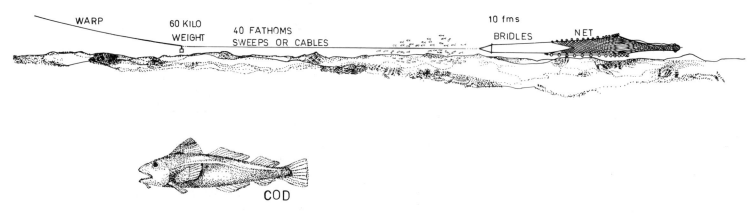

WARP   60 KILO WEIGHT   40 FATHOMS SWEEPS OR CABLES   10 fms   BRIDLES   NET

COD

Plates 8a and 8b   Contrast in midwater pair trawlers

Plate 8a   *Accord* PD90 steel trawler

Plate 8b   *Sparkling Star* traditional wooden vessel

usually carry fuel and water sufficient for three weeks' operation. Crews are small, only three or four men plus the skipper, though some vessels may carry an extra apprentice or cook.

Pair Trawl Nets

Cod pair trawls are made of nylon and have large meshes in the wings and square, usually 8 or 12-inch (20 or 30 mm) stretched length. For 150 hp vessels the net may be 400 meshes around the mouth and have a headline and footrope of 140 and 166 feet respectively (42 and 50 metres). Despite their size these nets are made of much finer twine than deep sea trawls, to give them a better 'flow' through the water. The meshes down the bag which may be made of polyethylene reduce gradually to 3 or 3½ inches (75–90mm). The whole net is of two-seam design with split or V-type wing ends to increase headline height. The bobbins are rigged on a ⅜ or ⁷⁄₁₆-inch diameter wire (10 mm) which is made slightly shorter than the footrope to make sure that the net does not come into contact with the sea-bed. A number of 16-inch (400 mm) diameter bobbins are placed around the bosom; usually from seven to eleven of them are used. The 12-inch (300 mm) bobbins are placed along the bunt or shoulder and may extend all the way to the bridle if particularly rough grounds are being worked. From four to sixteen of these may be fitted to the size of net under discussion. Along the lower bridle, from the wing end to the dan leno, rubber discs 1½ inch (35 mm) diameter may be fitted to reduce chafing wear. These rubber discs are also inserted between the bobbins on the groundrope. For working less rugged grounds, the bobbins may extend only as far as the shoulders, and rubber discs are used from there on. Every 2 feet (60 cm) or less, a 6-inch (150 mm) diameter rubber disc is inserted. This rig is quite common on small bottom trawls. The headline carries around 36 x 8 inch (20 cm) floats. Two 10 fathoms bridles connect the wing-ends to the bridle pole or dan leno. The footrope and groundwire are connected to a pipe clamp (see Fig. 26) which in turn is attached to an adjusting chain. The fishermen can take up or let out some links in the chain to trim the net properly. A bull-nose bobbin precedes the adjusting chain to protect the wing-end. Ahead of the dan lenos are the bridles of combination rope. These may be 40 fathoms long. A 60-kilo weight is attached between the warp and the bridle, but this may be replaced with a chain-wrapped tyre when towing on soft bottom.

Operation

Pair trawling operations commence after the vessels have located schools of fish by echo-sounder. Blind trawling is rarely attempted. Vessels may search the grounds for days before setting the trawl. Once fish have been detected and their position noted, the shooting vessel lies starboard side to windward and pays out the net. Once the net is over the side and clear of any foul-up, the second vessel approaches on the lee side and receives her bridle end attached to a heaving line. This is made fast to the starboard warp. The shooting vessel always takes the 'port' side of the net and runs its warp through the after side gallows. The second vessel pays the starboard warp out through the forward gallows and uses a towing block aft. Both vessels stream out the gear then drop the bumper weights and proceed to pay out the warp at full speed until they are about a quarter of a mile apart. Normally 350 fathoms of warp are used. When there are only 25 fathoms to go, the boats slow down and apply the winch brakes lightly to reduce the sudden momentum on the gear when the winch is stopped. Towing commences at 2¾ knots with the skippers maintaining contact by radio-telephone. Depending on the grounds and the fish marks, a tow may last anything from 15 minutes to 3 hours. They are usually much shorter than tows by deep sea trawlers. Some pair-skippers prefer to tow into the tide or wind to prevent excessive damage if the net is snagged. Before hauling commences, the boats draw closer and increase the towing speed to drive the fish well into the bag of the net. The closing of the wires is similar to the closing of the gear on a seine netter. Engine speed is reduced and hauling commences with the starboard vessel having knocked its warp out of the towing block. Both boats come slowly round to starboard while hauling. When the end of the warp with the bumper weight is up to the gallows the winches are stopped and the starboard cable end is passed to the port vessel by a heaving line. The hauling vessel then pulls in the cables, bridles and net. Some tows may result in over ten tons of fish and good strong lazy lines or choker ropes are needed to handle the bag, particularly in bad weather. The power block is also a great help when hauling in the gear.

The pair trawl gear is constructed in a way that enables the crew to make rapid repairs in the event of net damage. The ground rope is made up in sections and instead of being lashed to the footrope, the connecting chains are attached with 'rapide' clips. Similarly, screw-on floats have proved to be useful when a quick change of gear is required.

In order to check the fishing performance of pair trawl gear, the skippers check the warps and groundrope after each haul. They like to see the last 75 fathoms of warp fairly well 'polished' by the sea bed. Chains attached at the lower wing-end and around the centre bobbin(s) are also inspected for 'shine' to ensure that the net is on the bottom. The wing-end chains also serve as adjustors – links being taken up or let out to raise or lower the net and obtain satisfactory headline height. During some experiments with the gear it was found that headline height was normally around 12 feet, and increased to 14 feet when the boats were one eighth of a mile apart. Larger 400 hp vessels are now using the cod pair trawl successfully. They can probably operate in rougher sea conditions but it is doubtful whether, with increasing fuel costs, a more economical unit can be found than the 60-foot 170 hp Danish-style boats.

Fig 26   Bridle and cable gear: cod pair trawl

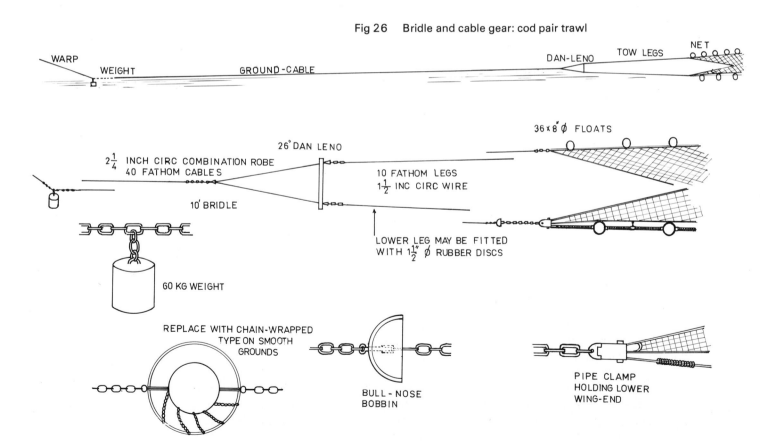

Modified to take Herring

The Danish cod pair trawl was successfully modified in Ireland in 1973 to act as a combination net catching herring as well as cod and haddock. Demersal species are relatively scarce on the West Irish grounds and herring is the mainstay of the trawler fleet. In the wintertime they use midwater pair trawls, but for most of the year, bottom herring trawls are used with otter boards. These nets take herring, mackerel and haddock in depths of from 30 to 80 fathoms during daylight hours. The herring bottom trawls were 'clean bottom' nets and could be extensively damaged if pulled over the rough grounds of which there are plenty off NW Ireland. Since much of the better quality herring and whitefish were found on the hard bottom, it was thought that a pair trawl similar to that used by the Danes, but designed to capture herring as well as cod and haddock, might be applicable.

For these experiments, the larger Danish pair

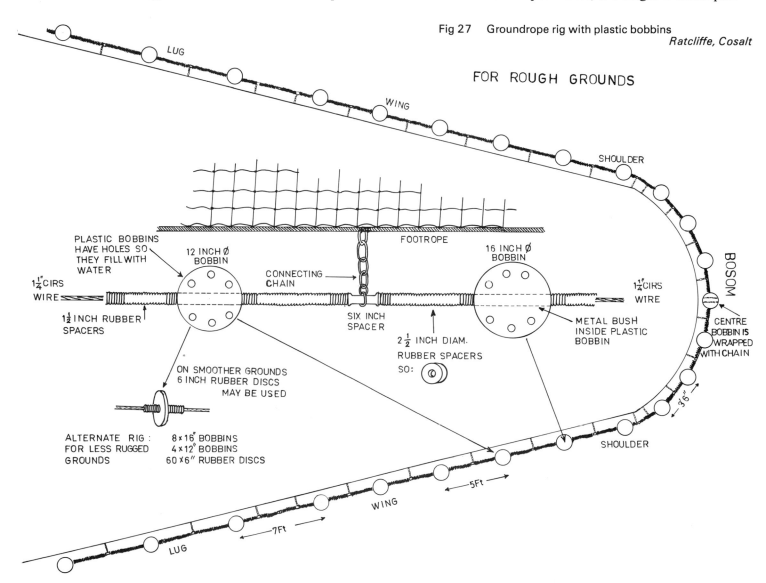

Fig 27   Groundrope rig with plastic bobbins

Ratcliffe, Cosalt

FOR ROUGH GROUNDS

LUG

WING

SHOULDER

PLASTIC BOBBINS HAVE HOLES SO THEY FILL WITH WATER

12 INCH Ø BOBBIN

FOOTROPE

16 INCH Ø BOBBIN

BOSOM

CONNECTING CHAIN

1¼" CIRS WIRE

1½ INCH RUBBER SPACERS

SIX INCH SPACER

2½ INCH DIAM. RUBBER SPACERS SO:

1¼" CIRS WIRE

METAL BUSH INSIDE PLASTIC BOBBIN

CENTRE BOBBIN IS WRAPPED WITH CHAIN

ON SMOOTHER GROUNDS 6 INCH RUBBER DISCS MAY BE USED

36"

SHOULDER

ALTERNATE RIG: FOR LESS RUGGED GROUNDS

8 × 16" BOBBINS
4 × 12" BOBBINS
60 × 6" RUBBER DISCS

5 Ft

WING

7 Ft

LUG

53

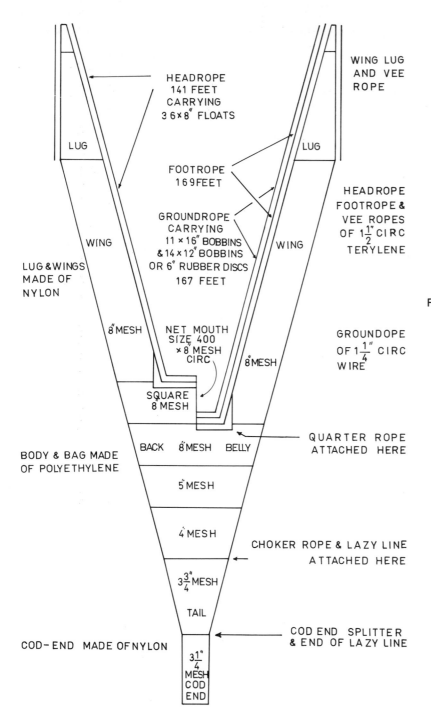

HEADROPE
141 FEET
CARRYING
3 6×8″ FLOATS

WING LUG
AND VEE
ROPE

LUG

LUG

FOOTROPE
169 FEET

HEADROPE
FOOTROPE &
VEE ROPES
OF $1\frac{1}{2}$″ CIRC
TERYLENE

GROUNDROPE
CARRYING
11 × 16″ BOBBINS
& 14 × 12″ BOBBINS
OR 6″ RUBBER DISCS
167 FEET

WING

WING

LUG & WINGS
MADE OF
NYLON

8″ MESH

NET MOUTH
SIZE 400
× 8 MESH
CIRC

8″ MESH

GROUNDOPE
OF $1\frac{1}{4}$″ CIRC
WIRE

SQUARE
8 MESH

QUARTER ROPE
ATTACHED HERE

BACK 8″ MESH   BELLY

BODY & BAG MADE
OF POLYETHYLENE

5″ MESH

4″ MESH

CHOKER ROPE & LAZY LINE
ATTACHED HERE

$3\frac{3}{4}$″ MESH

TAIL

COD END SPLITTER
& END OF LAZY LINE

COD-END MADE OF NYLON

$3\frac{1}{4}$″
MESH
COD
END

Fig 28   Typical 45 fathom cod pair trawl

(Number refers to theoretical
stretched mouth circumference)

trawl was selected. This type of gear used by vessels in the port of Esbjerg, is suitable for pairs of trawlers with from 300 to 450 hp each. The nets used were similar to the cod trawls except that they were fitted with small-meshed herring bags or 'brailers' 72 feet long. One large pair trawl (800 hp size) was used without bobbins; instead it carried 115 lb. (50 kilos) weight on the groundrope, and 23 x 11 inch (280 mm) diameter floats on the headline. A smaller 600 hp net was fitted with 56 plastic bobbins of 12 inch and 16 inch diameter (300 and 400 mm). The groundwire carrying the bobbins was 24 inches (600 mm) shorter than the footrope to make sure that the net did not come into contact with the bottom. The trawls were of two-seam design and had headline and footrope lengths of 55 and 66 metres respectively (180 and 216 feet). Two 75 foot 350 hp trawlers were paired up for the trials. These were of conventional side-trawler layout.

The tests were immediately successful, and proved that the bobbin pair trawl could be as effective on herring as it was on cod. The first tow with the large

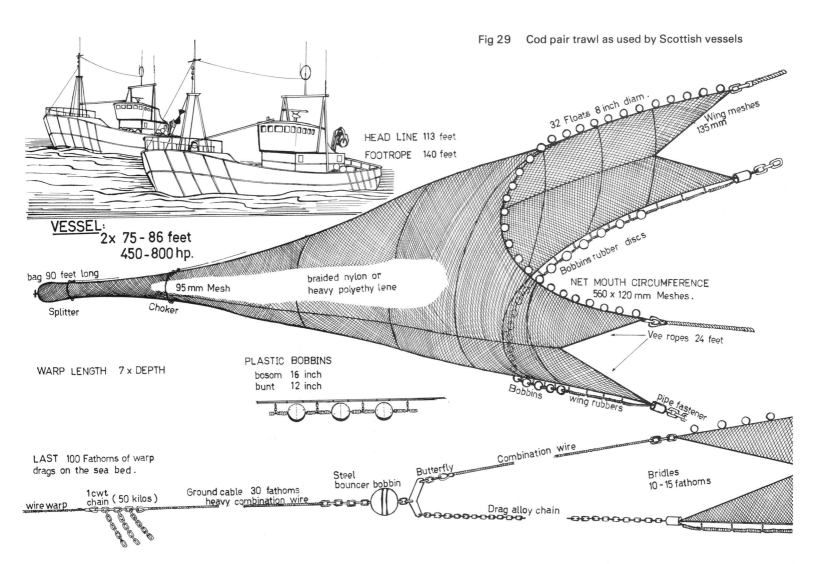

Fig 29   Cod pair trawl as used by Scottish vessels

HEAD LINE 113 feet
FOOTROPE 140 feet

32 Floats 8 inch diam.

Wing meshes 135 mm

VESSEL: 2 x 75 - 86 feet 450 - 800 hp.

bag 90 feet long

braided nylon or heavy polyethy lene

95 mm Mesh

Splitter   Choker

Bobbins rubber discs

NET MOUTH CIRCUMFERENCE 560 x 120 mm Meshes.

Vee ropes 24 feet

WARP LENGTH 7 x DEPTH

PLASTIC BOBBINS
bosom   16 inch
bunt   12 inch

Bobbins   wing rubbers   pipe fastener

Combination wire

Bridles 10 - 15 fathoms

LAST 100 Fathoms of warp drags on the sea bed.

Steel bouncer bobbin   Butterfly

wire warp   1 cwt (50 kilos) chain   Ground cable 30 fathoms heavy combination wire   Drag alloy chain

55

net on clean bottom resulted in so much fish, the bag burst. It is estimated that there were 800 boxes of fish in the net (40 tons). This was composed of 80% herring and 20% mackerel as estimated from the 10 tons that were retained. The large net later caught 280 boxes in a 40 minute tow in an area where a bottom trawler caught 120 boxes in a three hour tow. Similar results were obtained in further tests with the gear outfishing otter trawls by margins of 3 : 1 to 6 : 1.

The bobbin-rigged net appeared to fish closer to the bottom than the larger one with the lighter groundrope, for it took a greater percentage of demersal fish. From a three hour tow on rough grounds it took 180 boxes of fish which included 155 of herring and 25 of haddock. Despite the rugged nature of the grounds, the net was not torn.

The 'cod' pair trawl can therefore be used to good effect for herring and mackerel, with a minimum of modification. As with the cod gear, the herring bobbin trawl appears to be more than twice as productive as single vessel bottom herring trawls, and, of course, it can be used in places where the otter trawl can not, owing to the rugged nature of the sea bed.

Plate 9   Midwater pair trawling with small vessels. Clyde ringers engaged in pair trawling: *Pathfinder* BA 252 and *Ocean Gem* BA 265. Note the warp end weights hanging from the after gallows   *G Webster*

Development in Grimsby

At Grimsby, the 'home port' of cod pair trawling in Britain, there are two families of Danish origin who have featured prominently in the growth of this fishery. These are the Bojen and the Borum families. Jens and Jorgen Bojen began pair trawl operations in their little Danish seiners *Island* and *Sonia Jane*. The *Island* measured 57 x 17 x 8 feet and had an engine of a little over 100 hp. The *Sonia Jane* was around the same size, 34 tons registered, and had a slightly bigger engine. The rigours of the North Sea weather did not frighten the Bojen skippers at all, despite the small size and power of their craft. It was in little vessels like these that Danish fishermen had been operating in the North Sea for the past fifty years. Now they were to accomplish something that they could not do with the Danish seine net – they were to tow their gear over some of the most rugged grounds where cod shoals were to be found. And they were to succeed remarkably by mastering the skills of two-boat fishing in all weathers, by day and by night. Some of the catches and grossings by the Bojen skippers would have been astonishing for any size of fishing vessel. In August 1972 they landed 914 kits (57 tons) of prime fish for over £12,000. And this was from a five day trip. Later a newer vessel, the *Frances Bojen,* replaced the *Island* on the team, and in 1975 Jens Bojen paired up with another young skipper, John (Lemon) Richardson who commanded the vessel *Skanderborg*. The *Frances Bojen/Skanderborg* partnership had an amazing success in 1975, grossing over £140,000 in the last six months alone. Figures by themselves cannot tell the whole story, as with inflation money values are only relative. Suffice it to say that the figures of pair trawl catches and earnings shown on pages 156 and 157, are ones that deep sea trawlers of twice the length and five times the tonnage, would be pleased to have in 1975. On several trips, pair teams averaged over £2,000 per day at sea. During the same year, only the largest and most modern deep sea trawlers could surpass this. In these days of rising fuel costs, the production per horsepower of the pair trawl teams must surely be among the highest, and therefore the cheapest, in terms of both operational and capital costs.

By the end of 1976, cod pair trawl earnings had risen astronomically. Catches from single trips began to gross over £20,000 and over £30,000. The Bojen-Richardson team brought in one catch of 1,430 kits (90.8 tonnes) to gross over £40,000 for 13 days at sea. No doubt as money values decline these figures will lose their meaning, but for comparative purposes we might mention that in the same year, the largest and most successful distant water trawlers in Britain were averaging around £38,000 per trip, and around £1,750 per day at sea. Those vessels were three times the length, fifteen times the tonnage and had about seven times the power of the little cod pair trawlers. In terms of capital cost, one could probably build ten 65 foot, 280 hp pair trawlers (five pairs) for the cost of one 195 foot, 1,960 hp distant water trawler.

The *Frances Bojen* BCK 51 was built in the Jones Buckie shipyard in 1970. She is of typical Danish seiner layout, is only 66 feet long, and is powered by a Gardner 230 hp engine. A Hydema triple sheave pedestal power block was installed on the deck aft, and later a Lossie Hydraulics net drum was fitted amidships. Her partner vessel the *Skanderborg* H 64 was also built in Buckie, but is some 12 years older. She measures 63 x 18 x 8 feet and has a Gardner 152 hp engine.

In 1976 the *Frances Bojen* was teamed with a new vessel, the *Margrethe Bojen* BCK 11 which was also built at the Buckie Jones yard. Built of larch and oak along traditional Scottish lines, she measured 60 x 20 x 10 feet and was propelled by a 280 hp Kelvin engine with a controllable pitch propeller. On deck was installed a Norwinch TC8 9 ton 600 fathom trawl winch, and two Lossie Hydraulics pedestal type power blocks. The vessel carried 2,600 gallons of fuel (about 10 tons) and could accommodate over 42 tons of fish plus ice in her two holds. For fish finding purposes she was fitted with two Kelvin Hughes echo-sounders operating on 48 kHz and 28 kHz frequencies respectively, and a Simrad Sonar and scope. Other equipment included Sailor SSB and vhf radios, Atlas radar, Ben Amphitrite log and Decca Navigator MK 21. During her first eight months of operation, this compact modern vessel put ashore catches worth over £200,000 which probably exceeded her own total cost.

Plates 10a and 10b
Grimsby cod pair trawlers

Plate 10a
Margrethe Bojen

Plate 10b *Samantha*

More recent vessels which have been built with cod-trawling chiefly in mind, have had larger engines installed. Whether this investment in greater power is really worthwhile in pair fishing, remains to be seen. There are other 'multi-purpose' considerations, however, and if the vessels were also to engage in midwater pair trawling, then the extra power would be useful.

The Borum family, who formerly operated the *Solveig Borum* and the *Edith Borum,* 45 ton vessels with 152 and 200 hp engines respectively, now pair trawl with the *Carl Borum* and the *Jacqueline Borum.* Both vessels were built in Denmark on traditional anchor-seiner lines, but with modern equipment and layout suitable for cod pair trawling. The *Carl Borum* GY 322, which was delivered in 1975, is only 62.5 feet in length, but is powered by an Alpha 400 hp engine. She is equipped with a Norlau hydraulic winch and net drum. In her first month in operation she landed fish catches realising nearly £20,000.

A complete contrast in cod pair trawlers is provided by the *Mohave/Shawnee* partnership. These vessels are steel stern trawler types. They measure over 74 feet in length, have engines of nearly 500 hp and are equipped with hydraulic winches, net drums and power blocks. They are designed to engage in midwater pair trawling, seine-netting and bottom trawling as required. Most of the cod pair trawlers in Grimsby engage in Danish seining part-time, though some of them go bottom trawling on their own.

In 1974 there were seven pairs of cod trawlers fishing regularly from Grimsby. By 1975 the number had increased to eleven pairs or twenty-two vessels. This was considered enough for one port by the skippers themselves. Other pairs operate occasionally from Hull and from North Shields. A list of some typical vessels, with their equipment, is given in the table on page 150.

Scottish Use of the Cod Pair Trawl
Scottish fishermen showed considerable interest in the cod pair trawl, particularly the larger gear used by bigger vessels in Denmark, but it was not until the spring of 1976 that they began to use the gear

commercially. Herring trawling was then being restricted and with so many vessels reverting to seine-netting, there was some overcrowding on the fishing grounds. So a number of boats began pair trawling for cod and coalfish on the more rugged grounds of the North Sea from Aberdeen to Shetland. Fraserburgh and Peterhead vessels were among the first to adopt the gear. Though some of the boats were similar in size and power to the Grimsby pair trawlers, most of them were larger – in the 75 to 87 foot range, with from 460 to 850 hp engines.

The technique of pair trawling was not new to these vessels as they had been midwater pair trawling for years. Some of them retained the technique of using two warps from each boat, as with the midwater trawl. This worked well with the high opening lighter rigged pair trawls which were used on sandy bottom. With the heavier bobbin-rigged trawls, however, the single warp arrangement was used. Fifteen fathom heavy bridles were used with thirty fathom sweeps, and the warp length of around seven times the depth of water. About one hundred fathoms of warp, next to the sweeps, maintained contact with the sea bed thus 'herding' the fish in towards the trawl. Towing time usually varied from two to four hours, and the boats would tow about a quarter of a mile apart, depending on the warp length.

Pairs like the *Fairweather V* PD157 and *Sparkling Star* PD137; *Seringa* PD95 and *Sundari* PD93; *Unity* PD209 and *Morning Dawn* PD195; began to land catches of around 1,000 boxes (44.5 tonnes) from trips of one week or less. One pair, the *Faithful 11* PD67 and the *Ugievale 11* PD105, grossed £15,800 from a catch of 1,200 boxes taken on a five day trip. This averages out at over £1,500 per boat, per day at sea, which is more than some of the largest deep sea trawlers are able to earn. The catches consisted mainly of cod and coalfish. 'Coley' as the latter are called used to be unpopular on the market, but by this time they were selling for around £9 a box (= £20 per 100 kilos). By the summer of that year, more than 26 NE Scottish vessels were using the bottom pair trawl and more fishermen were showing interest.

The nets used by the Scottish boats were almost all of rectangular design, made of braided nylon or heavy courlene (polyethylene). The nylon nets were usually black dipped. They had 140 foot groundropes and headlines of 112 to 120 feet in length depending on the size of the square or overhang. The smooth bottom nets had meshes as large as 32 inch (800 mm) in the wings, tapering to 16 inch (400 mm) and 8 inch (200 mm) at the shoulders and mouth, and 4 inch (100 mm) in the bag. The heavier trawls had 135 mm wings (5.5 inch), 120 mm (4.75 inch) around the mouth and 95 mm (3.75 inch) in the bag. They were rigged with 16 inch plastic bobbins at the bosom, 12 inch ones along the bunt, and rubber discs at the wing ends. A steel bouncer bobbin and butterfly spreader connected the spreader wires to the groundcable sweeps. Companies making the trawls included Cosalt and Boris, England; Christensens, Denmark; Apeldoornse, Holland; and Syversens, Sweden.

As the Scottish trawlers were rigged for both side and stern trawling, and had large power blocks or net drums, setting and hauling presented no problems. The technique differed little from that used when they were midwater pair trawling. Unlike the Grimsby boats, they boxed all their fish, and this necessitated fairly large hold capacity. Fishing trips were relatively shorter, however, as most of the Scottish fishermen preferred to have their weekends at home. When fishing near the Shetlands some pairs would spend alternate weekends in Lerwick.

The bottom pair trawl also proved effective off the west coast of Scotland, on one of the largest remaining underfished species — blue whiting. These deep water fish are found at the edge of the Atlantic shelf from Norway to Ireland, but have not previously been fished to any great degree. Two 100 foot 750 hp Peterhead purse seiners, the *Vigilant* PD165 and the *Lunar Bow* PD118, spent five weeks pair trawling in up to 200 fathoms depth, from St. Kilda to the Faroe Isles. During that period they caught over 1,000 tons of blue whiting. The fish were kept in good condition in the refrigerated sea water tanks normally used for herring. In the absence of an established market for human consumption the fish were sold for pet food, but even at fishmeal or pet food prices (£50-£80 per ton) catches of 200 tons a week meant a good earning for these larger vessels.

Even larger vessels adopted the cod pair trawl in 1977. The J. Marr & Son trawling company of Fleetwood decided to try pair fishing with two of their large stern trawlers, the *Armana* and the *Navena*. These vessels were 130 foot, 390 ton, 1,700 hp deep sea trawlers. They were assigned the company's two top skippers, Willie Taylor and Victor Buschini. Operating on grounds off west Scotland and the Faroe Isles, the pair soon mastered the techniques successfully. Catches of over 2,000 kits worth around £60,000 were soon being landed from trips of just over two weeks duration.

5 Canadian Pair Seining

The same year in which Danish fishermen began pair trawling for cod in Grimsby (1969), fishermen 2,400 miles away in the Canadian Maritimes were trying out a two-boat method that was to be as successful in Canada as the Danish gear was in England. The boats used in Canada were even smaller than the pair trawlers in Grimsby.

Hundreds of small fishing boats operated from little ports in Newfoundland, Nova Scotia, Prince Edward Island, New Brunswick and Quebec. They were lightly powered and worked only traps, lines and gill nets in the waters of the Gulf of St. Lawrence and in bays opening into the Atlantic. Catches from long line gear were so poor that the vessels could barely pay their way. Lobster potting and salmon drifting were profitable methods but the season was very short. For the rest of the year the vessels lay idle or eked out a bare living using long lines.

It was primarily to provide these boats with a means of catching fish in the off-season that the Canadian government initiated a series of trials with new gear. The idea came from a former seine net skipper from Scotland who thought a kind of pair seine could be worked on a similar principle to the single-boat Scottish seine net. The gear was assembled and after preliminary trials and modifications, a satisfactory technique was developed. The results were immediately successful and with further experience they improved even more. The two little boats (38 feet, 112 hp) came ashore day after day loaded with fish, sometimes 3, 4 or 5 tons of flats and hake. On one day they landed over 20,000 lb. of fish, mostly hake. Over the project test period of 37 fishing days they landed 152 tons of fish, an average of over 4 tons (90 boxes) a day. The following season, 14 boats adopted the new gear. This number was doubled in two years, and Canadian pair seining was born. It spread to all of the Maritime Provinces and to the Canadian west coast. In 1972 a pair of 40 foot boats from Georgetown Prince Edward Island landed over a million pounds of fish (500 short tons, or 10,000 boxes). This is equivalent to what many 80 or 90 foot trawlers land in a year.

By 1973 the gear was being tested on both the east and west coasts of America, and modifications were being made to enable it to be used in freshwater lakes.

The pair seine operation consists of sweeping an area of the sea bed rather than actually towing a trawl. The net does move about 1½ or 2 miles during an hour-long set, but it is the spread of the warps which herd the fish into the net that makes the gear so successful. It must also be borne in mind that these small boats do not have the power to tow a trawl. They only really 'tow' the net at the critical closing stage when the boats draw close together before hauling in the warps. In order to maintain the remarkable spread of the gear, wire warps are used on each side of the 'V' type set (see Fig. 30). The wire digs into the soft bottom and keeps the gear open. Because the wire grips the sea bed in this way, it cannot be used next to the net where the warp is set out at right angles to the direction of tow. Manila seine-net ropes are used here as they come over the bottom easily. The net used is a light wing trawl or bottom seine and the gear appears to fish best with a small net – a 320 wing trawl or a number 6 seine net.

The vessels used for pair seining are typical of the 35-45 foot Canadian lobster or scallop boats. These craft are transom sterned with a small wheelhouse forward. They usually have around 100-150 hp high speed diesel engines. (The 150 hp engines used by Grimsby pair trawlers are heavy duty slow-speed engines.) The boats carry a mast and boom or derrick just behind the wheelhouse. They are light craft with limited freeboard. Loaded up with 5 tons of fish, a 38 foot boat would have only 15 inches of freeboard remaining. The chief modification necessary for pair seining is the installation or

Plates 11a–d Stages of Canadian pair seining operations   *Dept. of Fisheries Canada*

Plate 11a   Shooting: note the winches warps and tyre fenders

Plate 11b   Towing

Plate 11c   Hauling

Plate 11d   Hauling nearing completion

Plate 11e Taking the catch on board

Plate 11f The catch consists mainly of soles, witches, red hake and cod

modification of the winch which should be able to accommodate 400 fathoms of ⅜ inch diameter wire, and 240 fathoms of 2⅜ circumference manila rope. Holes are drilled in the rail aft for the towing rollers which should be moved to either side depending on the direction of the tide. Old car tyres are lashed to one side to act as fenders when coming alongside the partner boat in fine weather.

The dimensions of the boats used on the east and west coast of Canada, for the initial trials were:

Pair Seine Vessels: Canada

	Prince Edward Island		British Columbia	
	Vessel 1	Vessel 2	Vessel 1	Vessel 2
	Norma M	*Mary Lou II*	*Gambler I*	*Taplow II*
L x B x D in feet	38.5 x 12.2 x 3.7	40.4 x 12.1 x 3.8	38 x 12 x 5	45 x 12 x 6
Mast	18' x 7" diam.	18' x 7" diam.	20' x 10" diam.	20' x 10" diam.
Boom	16'	16'	20'	22'
Fish hold capacity	5 short tons	5 short tons	10 short tons	12.5 short tons
Engine	112 hp Perkins	112 hp Perkins	150 hp Volvo Penta	110 hp GM diesel
Propeller (3 blade)	20' x 24"	20' x 24"	30' x 24"	36' x 22"
Reduction gear	2.5 : 1	2.5 : 1	2 : 1	3 : 1
Power take-off	Ford 4-speed gearbox	Ford 4-speed gearbox	Twin disc	Twin disc
Electronics	Radio and echo sounder	Radio and echo sounder	RT Radio, echo sounder Radar and Loran	RT Radio, echo sounder Radar and Loran

All of these vessels were fitted with a modified winch drum having large (30 inch diameter) flanges. This enabled them to hold 400 fathoms of ⅜ inch diameter wire and 240 fathoms of 2⅜ inch circumference manila rope. The British Columbia vessels also used a net drum which was handy for hauling in the gear. In the first trials it was found that the 112 hp vessels had adequate power to tow the gear. They towed comfortably at 1,200 rpm which was only a little more than 50% of maximum power (2,200 rpm). The cost of equipping the boats with a winch, warp, ropes, towing roller, net and

63

spare twine was less than $2,000 each in 1969. In view of the vastly increased earnings this was considered to be a relatively small investment. The British Columbia boats which used the gear in 1972 did not require modified winches. The cost for them of warps, ropes, towing rollers and nets, was only $1,500 per vessel. It should be noted that an echo sounder is essential for this kind of fishing and as many lobster boats are not equipped with one, the price of a small sounder, plus installation should be added to the cost of outfitting for the pair seine. Canadian pair seiners use either a small wing trawl or small seine net. These two nets appear to fish equally well with the gear. The wing trawl may take larger quantities of round fish, particularly in deep water, but it is prone to distortion if it is not towed directly into or with the tide. For this reason, in areas where the current varies a lot in direction, fishermen prefer to use the seine net. If using a wing trawl, they set a marker buoy to a light anchor to indicate the direction of the tide. A 320 wing trawl is probably the most popular size of net. The number 320 refers to the meshes around the mouth (usually 4 inch mesh size). This would probably give an actual net mouth circumference of 50 feet when fishing, in an oval shape measuring about 10 feet high by 16 feet wide. Cod-end mesh sizes used in Canada and the USA are slightly larger than the minimum size used by inshore vessels in Europe when working for white fish (70 mm stretched).

The net is connected to a simple wooden dan leno or bridle pole by 5 or 10 fathom lengths of combination rope about ½ inch diameter. The manila warp is shackled to the bridle which has a swivel attached to the G link. Another swivel and shackle link connects the manila rope to the trawl wire. On deck the warps pass through a 'seine-net' towing roller which consists of one horizontal roller and two vertical guide rollers. The roller, which weighs about 40 lb., can be shifted to any of three or five holes in the rail aft, to provide a different towing position. The boats carry a crew of three men – one skipper and two deckhands. This is sufficient for handling the gear, but a total of four men would be better, particularly when large catches of fish must be handled. The deckhands from the second vessel

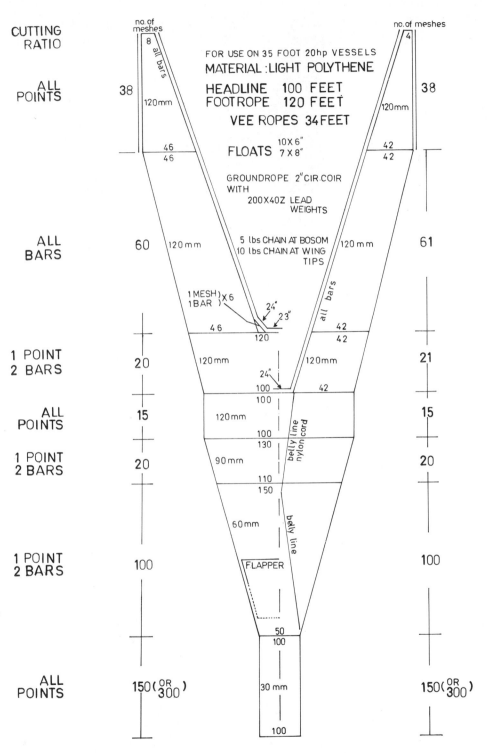

Fig 30   Pair seine trawl for 20 hp vessels   *L Innes*

64

can transfer to the hauling boat to assist in taking the net in – provided the weather is fine. In bad weather, it is not advisable for the boats to come alongside each other.

Operation

Each boat is equipped with a complete set of gear but one of the skippers is generally appointed as the 'commodore skipper' and he is in command of the operations regardless of whichever vessel is setting the gear. The fishing can be carried out without radio contact but it is much better if each vessel has a vhf radio telephone or a set of 'walkie-talkie' radios.

Once the boats have arrived at the grounds, the net is paid out over the stern of the shooting vessel. It goes over cod-end first as on a trawler, and not wing-first as on a seine netter. The second boat passes its manila warp end across and this is shackled on to the bridle. The bridle poles then go over the stern, and while gripping softly on the winch brakes, the boats spread apart in opposite directions at right angles to the direction of the tow. When the net wings are spread apart, the warps are released and paid out until the wire is reached. In shallow water or with very limited engine power, only one coil of manila rope is set (120 fathoms), but normally two coils of

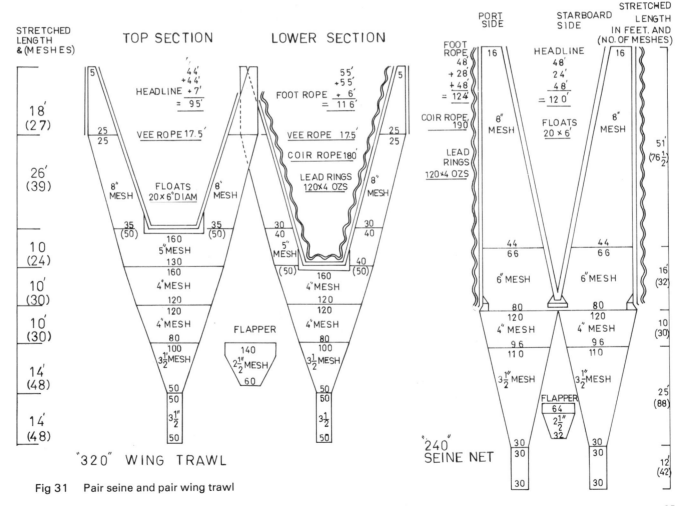

Fig 31   Pair seine and pair wing trawl

rope are set by each boat. This puts the vessels 2 x 240 fathoms (or 480 yards) apart. This, plus the net and bridle length is a quarter of a sea mile. The boats turn and head in the towing direction and begin paying out the wires. It is important that the vessel has turned before the wire is released. Any wire set across the direction of tow will dig into the sea bed and snag the gear. When 300 or 400 fathoms of wire (depending on the depth) have been set, the winch brakes are tightened and the vessel, which will have slowed right down, begins to tow at about half or three-quarter engine speed. The gear is nearly always towed with the tide. The boats tow ahead at about 1½ knots. This plus the effect of the current, will bring the gear over the sea bed at around 2 to 2½ knots.

The wires on each side dig into the sand and keep the gear open. The last 75 fathoms of wire should be 'polished' by the sea bed. The boats steer straight ahead, and no attempt is made to stay apart. After about one hour, the boats turn slightly towards each other and draw the wires together. As they approach they speed up a little to drive the fish into the bag. Once the wires are parallel, hauling commences with the propellers still driving ahead, but slowly now. If the sea is calm the boats can come alongside each other and make fast fore and aft. They are equipped with tyre fenders for this purpose. Should the sea be choppy, however, there is no need for the boats to make contact. Hauling can continue and be completed while they are one or two boat-lengths apart. Once the bridle poles are up, the second vessel passes its bridle to the hauling boat and the net is taken in by hand or by net winch if one is available. As much as 5 tons of fish (100 boxes) can be taken in a single one hour tow. The little boats take the catch aboard in lifts of a quarter to a half a ton. If the catch is good, both vessels steam back up-tide and repeat the operation.

Throughout the setting and hauling sequence the vessels must co-ordinate their actions. As they must tow on the same compass course, it is imperative that they have both compasses adjusted and one compensated if it is not reading exactly as the other. When hauling in the warp, care must be taken to keep it equal at all times. The wires and ropes are

Plate 12   Canadian pair seining. Taking the catch on board
Dept. of Fisheries Canada

marked at 25 fathom intervals and one winch must stop a few seconds if its mark comes in ahead of the other. If the gear is not set exactly with the tide, then the vessels must tow at an angle to the warp as illustrated in the sketch. This situation is where the extra holes for the towing rollers come in handy. By shifting the towing point towards the port or starboard side it is possible to tow at an angle and thus counteract the current. If there are no reliable land marks in the vicinity, a marker buoy can be set at the point where setting commences. This is a valuable guide to the vessels as the current may drive them much further than they realise. Smooth clean bottom is essential to pair seining. Unlike the cod pair trawl, it simply cannot be worked on rough grounds. Sand, sand and mud or sand and shingle are the best types of bottom, and the gear can be worked in depths up to 60 fathoms. The 40 foot 100 hp–120 hp vessel is probably the most economical size of unit and fuel costs are remarkably low, which is an important factor given present day fuel prices. The gear has been used successfully on 55 and 60 foot boats but it is primarily designed for the smaller vessel.

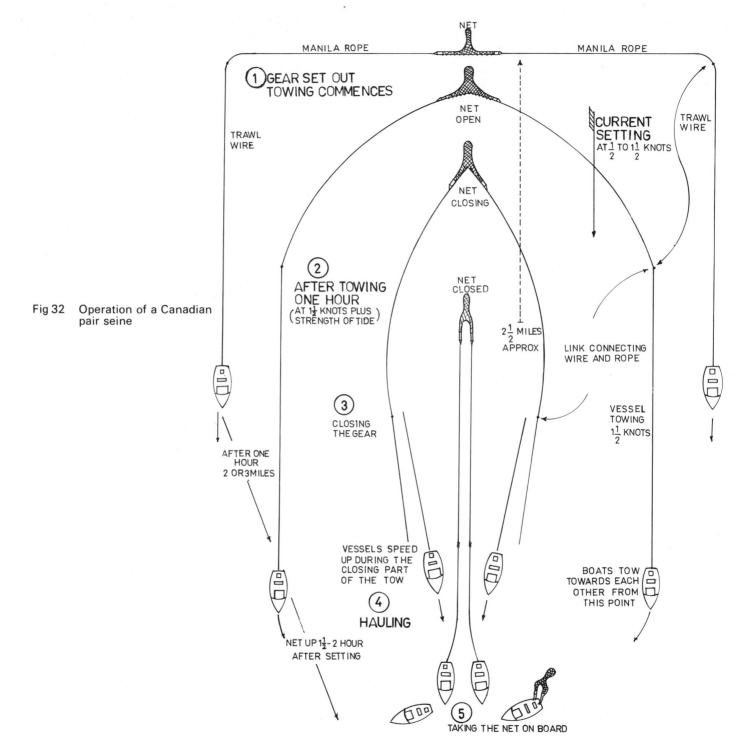

NET

MANILA ROPE   MANILA ROPE

① GEAR SET OUT
   TOWING COMMENCES

NET OPEN

TRAWL WIRE

CURRENT SETTING
AT $\frac{1}{2}$ TO $1\frac{1}{2}$ KNOTS

TRAWL WIRE

NET CLOSING

② AFTER TOWING ONE HOUR
(AT $1\frac{1}{2}$ KNOTS PLUS)
(STRENGTH OF TIDE)

Fig 32   Operation of a Canadian pair seine

NET CLOSED

$2\frac{1}{2}$ MILES APPROX

LINK CONNECTING WIRE AND ROPE

③ CLOSING THE GEAR

VESSEL TOWING $1\frac{1}{2}$ KNOTS

AFTER ONE HOUR 2 OR 3 MILES

VESSELS SPEED UP DURING THE CLOSING PART OF THE TOW

BOATS TOW TOWARDS EACH OTHER FROM THIS POINT

④ HAULING

NET UP $1\frac{1}{2}$ - 2 HOUR AFTER SETTING

⑤ TAKING THE NET ON BOARD

Success Factors

The success of pair seining is due to the size of the area of sea bed which is swept in a single tow. The gear covers an area of bottom of 400 fathoms (2 x 2 coils of manila rope) x 75 fathoms (the length of wire on the sea bed).

This covers an area of bottom 30,000 sq. fathoms. Multiply this by the distance towed in one hour at 1½ knots plus 1 knot tide, *ie* 2½ miles or 2,500 fathoms. This is 33 times 75 fathoms so the total area covered is 30,000 x 33 = 990,000 sq. fathoms.

Compare this with the ground covered by a modern deep sea trawl with 75 fathom cables between the net and the otter boards. If the distance between the otter boards is 60 fathoms (assuming maximum door spread) then the area covered is ½ x 60 x 75 which is 2,250 sq. fathoms. Multiply this by the distance covered in a three hour tow at 3 knots. Three by three gives 9 miles which is 9,000 fathoms or 120 x 75, so the total area covered is 2,250 x 120 = 270,000 sq. fathoms. We see, therefore, that the little pair seine boats, in one hour, cover more than three

Fig 33   Layout of a Canadian pair seine net

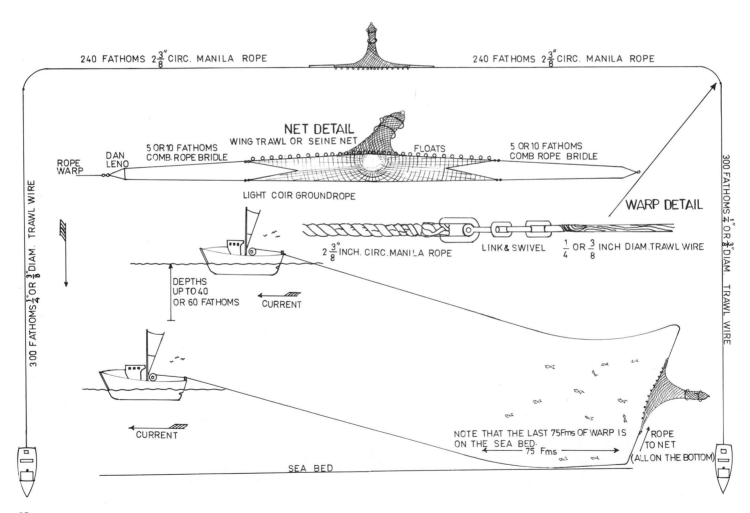

Plate 13   Preparing to come alongside   *Dept. of Fisheries Canada*

Pair Seine, Canadian West Coast Fishing Report MB *Gambler* and MB *Northern Breeze*

The best day's fishing had with the pair seine was in the Hecate Straits on Monday, 23rd July, 1973. We set in Lat. 54° 21'N Long. 131° 12'W at 11.30 a.m., set to tow E.N.E. through 41 to 43 fms. This was a kind of gulley, the water shoaling to 24 to 30 fathoms on either side. Three hundred fathoms of 2½ inch rope were set on either side of the net, and 300 fathoms of wire on either side on the line of tow.

We towed for one hour before closing the gear. Our tow was directly before the tide, and we covered the whole gulley in the 1 hour tow. We found whilst heaving up that the net was very heavy; I thought we had a rock or a tree in the bag, but the excessive weight was mainly English sole. However, there was a lump of clay or mud in the cod-end. We did not take the first lift onboard, opening the cod-end as it hung from the derrick and letting the fish and mud go back into the sea.

After that we took eight lifts of clean fish, 15,000 lb. in all. This was the most flat fish I ever saw in one set. I may add that the net was a new 420 x polyethylene Vinge Seine trawl with four heavy rib lines from the bosom to the cod-end. The weather was calm at the time. It took us four hours to get the fish out of the net. We put 10,000 lb. on to the *Northern Breeze* and 5,000 lb. on to the *Gambler*; we had 30,000 lb. for that day's fishing and landed 100,000 lb. for our five day trip, 20,000 lb. of cod amongst that. We had about 2,000 lb. of rock sole and 1,500 lb. of grey sole or witches.

We landed our catch at Prince Rupert, which was approximately four hours distant or 40 miles.

Capt. James C. Thomson

times the ground covered by the deep sea trawler in three hours. There are, of course, many other factors to be considered, but it would appear safe to aver that pair seining probably offers small vessels the best means of catching demersal fish on smooth grounds.

It is interesting to compare the Canadian pair seine with Japanese bull trawl gear. The former appears to be a miniature reproduction of the latter, although it was developed quite independently. Both gear types use a relatively large net for the type of vessel, and both rely on the sweeping ability of rope cables next to the net, and the bottom-gripping properties of wire warps at each side of the set. The two methods are vastly different in terms of vessel size and power, the one being suitable for deep sea trawlers, and the other for inshore 'lobster' boats.

6 Freshwater Pair Trawling

Many sea fishermen do not realise how much fish is taken from freshwater rivers and lakes. They can be of considerable commercial importance particularly in tropical and sub-tropical countries. A land-locked country like Zambia for instance produces nearly 50,000 tons of fish annually. This is not far short of the production by some of the smaller European countries. There are large freshwater fisheries in China and in the Soviet Union. It is estimated that over 10 million tons of fish annually are taken from inland waters throughout the world. This is about 15% of the total world catch.

The bulk of freshwater fishing is carried out with artisanal type fishing gear such as gill nets, traps, beach seines and hand lines. Purse seines and trawls have been used in the bigger lakes in Africa and Asia. Trials with two boat trawls have been successful in Europe and Canada and it appears likely that this technique will spread to freshwater fisheries in Africa, South America and Asia. The problem in most lake fisheries is the absence of adequately powered craft to tow movable gear. However, the developments in Germany and Canada have proved that small trawls can be successfully fished by a pair of boats with as little as 5 or 10 hp each. This shows considerable promise for canoe-type fisheries which employ only outboard engines, or the smallest types of inboard diesel engines.

German Experience

The German freshwater pair fishery began in 1964 when trials were conducted to improve the fishing methods. The growing shortage of manpower necessitated a more mechanised operation. The main species sought in this case was eel *(Anguilla anguilla)* which is an expensive and commercially important fish. A two-boat bottom trawl, similar to the Baltic pair trawl, was constructed. It had very small meshes, particularly in the cod end (12 mm)

and was small enough for vessels of 10-20 hp to use. The headline and footropes measured only 40 and 45 feet approximately. The gear was lightly rigged with floats and weights, and towing warps of only 3 mm diameter wire or 10 mm diameter rope were used. A hand operated winch was used in the case of the wire warps. The two 12 hp vessels were able to tow the gear at 1½ knots which was reckoned to be sufficient. A warp length of 100 metres was used in depths up to 10 metres. At the warp end was a 7 kg. weight followed by 8 fathom sweeplines, a 2 foot dan leno, and 6 fathom triple bridles. Only 8 kg. weight was used on the groundrope.

The eel pair trawl proved to be very successful in the shallow freshwater lakes with rather undulating bottom contours. It could be operated by two 8 or 10 metre boats with only one skilled man on each. Some of the boats were half-decked diesel powered vessels, but others were simple outboard powered skiffs. As the lake bottom was mostly soft mud there was no danger of tearing the net, only of 'mudding up' if the footrope was not properly rigged. Hauling was accomplished by hand with the vessels facing the net.

The German eel fishery extends from May to October. In order to lengthen the fishing season and take a greater variety of fish, a larger meshed net was designed to catch perch and pickarel species. It had 40 mm meshes in the wings and 25 mm meshes in the cod-ends. Otherwise it resembled the eel net in design and rigging. Both nets were conventional two seam trawls and were fitted with flappers at the cod-end mouth to prevent fish from escaping during the hauling period.

Canadian Experiments

The Canadian freshwater pair fishing had a later and slightly different origin. Canada, of course, has vast stretches of inland waterways and countless number of lakes, large and small, extending from latitude 42 degrees to inside the Arctic Circle. As the winter climate is so severe, these lakes are not important as commercial fisheries. They are, however, vital to sports fishermen, and to the national tourist industry. For this reason, it was deemed useful to harvest the unwanted species at the end of each season, and thus leave more food and room for the growth of the next year's crop of sports fish, mainly trout and salmon.

As with most freshwater lakes, there was a problem with the size and power of the boats available. Most of them were outboard powered skiffs and had no kind of mechanical winch. However, small sea-going craft were then having some success with the pair seine in the Maritimes, and it was thought that a smaller version of this gear might work in the lakes.

To make the net as easy to tow as possible, it was

Fig 34   Freshwater pair trawl
Steinberg, Modern Fishing Gear of the World 3

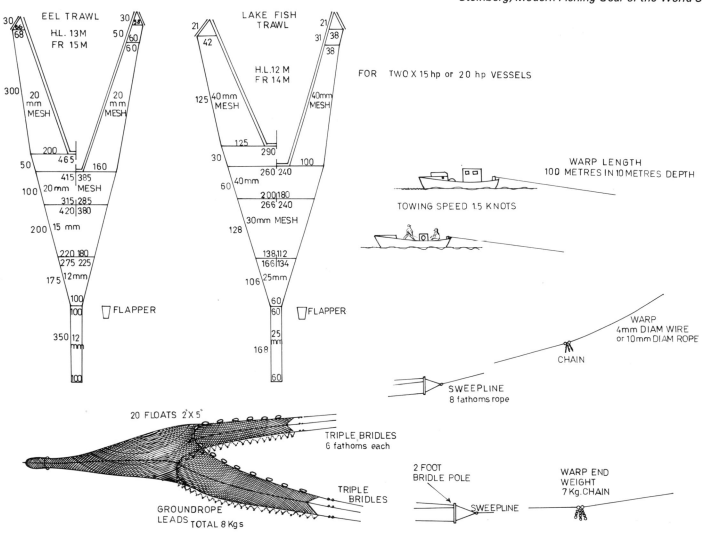

71

constructed entirely of knotless netting made of fine nylon twine. This was found to be ideal for the conditions and the gear was eventually used successfully by two 5 hp engined vessels. As much of the water was extremely shallow, sometimes less than 8 feet, a shallow net was required, and this was constructed on the lines of a bottom seine. For deeper lakes, a deep seine or miniature wing trawl was used. The gear was towed both on the pair seine principle, and as a straight bottom pair trawl.

The gear fished best in clear waters. When the water was muddy or cloudy, the 'herding' effect of the warp was not possible and it was just as useful to tow the net like a pair trawl. In some cases the fish concentrated in deep holes near where rivers or streams entered the lake. It was a simple matter to set the gear around these holes and net the fish in a short tow. In clear water, however, the herding effect worked well on the smooth shallow bottom. Sometimes the lake bed was covered with weed, but this did not interfere too much with the operation. In shallow water it was observed that if the floats 'broke' the surface at all, then no fish were caught.

The Canadian freshwater pair fishing was conducted mainly for lake herring and pickerel, but sometimes more valuable fish were caught. Trout were taken in good quantities in the knotless net. As the tows were of short duration the fish were taken in excellent condition. They were lifted aboard in small quantities or 'brailed' out of the net to avoid damaging them in any way.

The trout pair seine had a headline and footrope of 115 and 117 feet in length, respectively. It was made entirely of 210d/15 knotless nylon 1½ inch (40 mm) mesh, except for the cod end which was of 210/21 twine. The seine mouth had a circumference of 400 meshes which would result in a diameter of around seven or eight feet. The headline carried five 5 inch and twelve 2 x 5 inch floats. About 18½ lb. (8½ kilos) of weight was attached to the grass ground rope. A set of light bridles, dan leno and warps completed the gear which could be handled by two or three men.

Larger nets and heavier gear were used in lakes where more powerful boats were available. Some of these craft had gasoline inboard/outboard engines

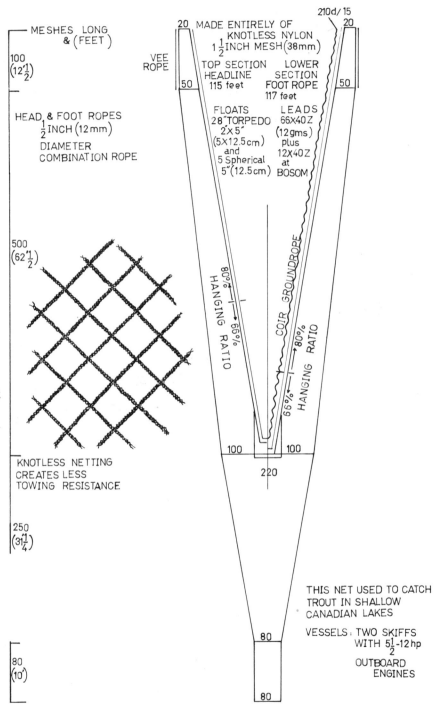

Fig 35   Freshwater pair seine

72

of around 100 hp. The results were usually the same. In clear waters, the gear fished excellently. No great towing speed was required, and as with the German lake trawling, a speed of around 1½ knots was found to be sufficient. The knotless netting flowed out into perfect shape as observed from the surface, and created less towing resistance than would be formed with knotted webbing. Catches of lake herring and other species were as high as 5,500 lb. (2,500 kilos) from tows of 30 to 40 minutes' duration.

The Canadian experiments were observed by African fishery officers who plan to carry out trials with similar gear in some of the more productive lakes in Africa where fish are sometimes 'dormant' or inactive which makes them difficult to catch in gill nets. A small freshwater pair trawl might even be pulled by two sail-powered fishing canoes. One of the significant findings of both Canadian and German freshwater pair trawling, was that a towing speed of only 1½ knots was adequate. This is an encouraging factor for lake fisheries with small, light powered fishing craft.

African Trials

In Lake Chilwa, Malawi, a bottom pair trawl was successfully operated from open boats with 5 hp outboard engines. The net, constructed of very fine nylon twine, 50 mm and 25 mm mesh, had a headline and footrope measuring 20 metres each. A safety line, 11 metres in length was attached between the wing ends to prevent damage or overstrain. Using 30 metre polypropylene warps in shallow water, the skiffs were able to tow the net successfully to capture carp, bream and other tropical fishes.

The success of bottom pair trawling in freshwater lakes indicates that it might be possible to use a midwater pair trawl in such waters, designed specially for lightly powered craft. In the deeper lakes of Africa and Asia, it may be possible to use a midwater net constructed entirely of light knotless nylon, which could be towed by two vessels with less than 50 hp. It could form the basis of a most economical mechanised fishery requiring only small investment.

The potential of midwater trawling in freshwater lakes might be indicated by results of experiments in Germany where pike and smelt were caught in a small midwater pair trawl of rectangular design. This net was towed by two 20 hp vessels using single warps to two spreader poles separating two pairs of 10 fathom bridles. Ten kilograms of weight were attached to each lower wing end. It was found that a net mouth opening of 50 m² or less was not really sufficient. In this case the mesh size used in the wings and mouth — 40 mm, was rather small. By using larger tightly hung meshes of fine nylon or polyester, it may be possible for vessels of around 20 hp to pull a trawl with a bigger net mouth, and thus achieve better results. Experiments in Europe, though moderately successful, are limited by the shallowness of the lakes and the lack of abundant stocks of fish. In the huge deep lakes of Africa and some parts of Asia, however, there are no such limitations, and the presence of large quantities of freshwater sardines and other shoaling species would appear to warrant trials with two-boat midwater trawls suitable for small boats. It would be better if such craft were equipped with winches. This makes it much easier to vary the net depth when towing. Hand-operated windlasses can be used in such situations, but of course their hauling speeds are slow. Echo sounders are indispensable to midwater fishing, and it is quite possible to equip a small open boat with a battery-operated paper-recording echometer. Several small portable units are available today at reasonable cost. The transducer is fixed to a pipe which can be bolted to the side of the boat, and may be unbolted easily before entering the harbour. The display cabinet is bolted to a little frame in a suitable location, and the whole unit may be removed and taken ashore after each trip. Unless some kind of fish attraction is used, midwater fishing without echo sounder is just not feasible. Trawl fishing in three dimensions demands accurate pin-pointing of the location and depth of the fish schools to be truly successful.

The single-warp method of midwater pair trawling for small vessels is described in more detail at the end of chapter seven. Figure 34 shows the arrangement of the gear on the German 15 hp vessels. Because of the limited power of the boats, it was safer to regulate the net depth by the length of lines connecting the warp-end weights to buoys floating on the surface. In the case of the lighter nets towed by small vessels, it is a wise precaution to use a bow line between the two boats to help them maintain the correct distance apart, and prevent damage to the net. The twine used in the freshwater trawls is extremely light – usually around 200 to 500 R tex, or 210/9 to 210/21 denier nylon.

Midwater trawls must be constructed and operated to a fair degree of precision to be successful. Whereas a poorly designed bottom trawl will always catch some fish, however few, it is

unlikely that a badly made midwater trawl will take any fish at all. It is indicative of the advance of fishing technology that it is now possible to make such nets light enough to be fished successfully by small open boats. Because of the limited power and size of these craft, it is important that the whole gear be simple to operate and easy to handle. The single-warp buoy-supported net of fine nylon fulfils these requirements. The two-boat towing technique makes it possible to use even the smallest of fishing craft for this purpose.

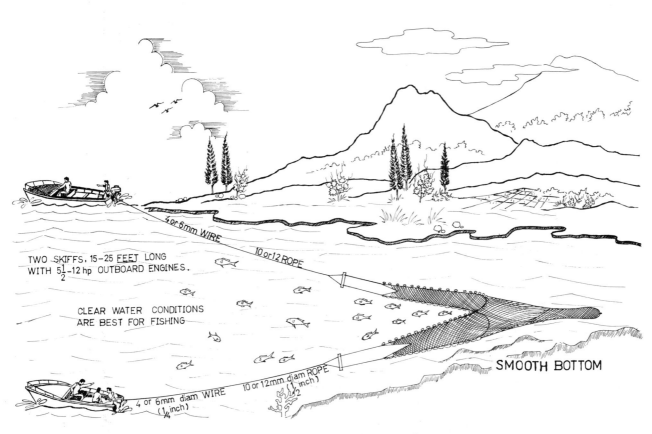

Fig 36   Freshwater pair seine operation

Fishing Report: Lake Nipissing, Ontario, Lat. 46° 20′N Long. 80°W

Before reporting on this one day's fishing with Canadian pair seine I would like to mention that this lake is very shallow – an average of approximately 7 fathoms or 42 feet in depth. Previous to this we tried pair trawling on fair to good fish marks on the bottom but the most fish we got for one hour of towing time was 1,200 lb. This was mostly sisco with about 50 lb. of white fish. We began to realise that the noise of our motors was affecting the fishing so we decided to change over to Canadian pair seine, and the difference in fishing had to be seen to be believed. We then caught more than four times the quantity of fish per set. The main reason for this can be attributed to the stealth of the gear or its silence, the net and ropes being half a mile distant from the vessels when towing.

In pair seining we depend on herding the fish we have encircled into the path of the net which is being towed slowly at about 2 miles per hour, whereas in pair trawling or any other trawling we depend on the fish in front of the net. In pair seining we use two coils per side of 2¾ inch circumference manila rope. This gives a length of approximately half a mile from the net at the start of the tow, that is half a mile from bight to bight of the set, and half a mile of wire on each side set parallel to each other, four coils or 400 fathoms apart and at right angles to the ropes of the sweep. That is, if we set the net and ropes east or west we set the wires north or south depending on which side of the net the bag is on. If the bag is on the north side of the net we tow south and vice versa.

We set out from Sturgeon Falls on the 1st July, 1974 with our two 30 foot boats fitted with 100 hp outboard motors, and steamed 1½ hours to the mouth of the French river. We set at 9 a.m. and towed for 30 minutes and got about 4,000 lb. of fish, that was 3,000 lb. of sisco (average size 1 lb. each), 700 lb. of white fish (average 2 lb. each), 300 lb. of ling cod (average 4 lb. each). We set again and got 3,000 lb. of the same quality as the first set. We had more fish for the two sets than we could box and returned to Sturgeon Falls loaded at 2 p.m. the same day. We had 3 weeks of fishing in this area and catches ranged from 3,000 to 7,000 lb.

Capt. James C. Thomson

Plate 14   Japanese GRP pair fishers: ideal for lake or coastal work
Yamaha Motor Co

Plate 15   Freshwater pair seining on Lake Nipissing
—setting the gear   *Tribune*

Plate 16   Hauling in the catch   *Tribune*

Plate 17   Taking the catch aboard   *Tribune*

7 Midwater Pair Trawling

Two-boat midwater trawling is by far the most important kind of pair fishing in the world today. Fleets of pair trawlers working mostly for herring, operate from ports in Britain, Denmark, Ireland, Sweden, Germany and Holland. The method is also used to a lesser degree in Canada and the USA. It began in 1943–44 when Yngves Berntsson of Foto Sweden made a successful midwater pair trawl. The gear was improved by Robert Larsen of Skagen Denmark who tested his 'Atom' trawl during 1945-48. He produced a four-seam design net which is in essence no different from the standard pair trawls now used all over Europe, a quarter century later. The gear fished very well in the Baltic and later in the North Sea, and it was soon adopted by fishermen in Sweden, Germany, Ireland and Holland. English sprat fishermen adopted the gear for their small vessels, but Scottish fleets were slow to take it up. There are now hundreds of pair trawlers in Scotland, but it was not until the mid 1960's that the method caught on there.

To understand the development of midwater trawling we must look at the history of the herring fisheries. The herring is by far the most important commercial fish in Europe. For hundreds of years it formed a vital part of the staple diet of people in countries bordering on the north east Atlantic. Salt herring could be stored for months without losing any of its excellent food value and it could be cured in barrels quickly without any need for the dry weather necessary to produce salt dried cod, ling or halibut. The salting and curing of herring was a highly skilled trade two hundred years ago.

Herring were caught in large quantities in drift nets which Dutch fishermen pioneered. The drift net was the chief means of catching herring until the mid-twentieth century when the purse seine boom started. During the period 1920-1940, however, the fishery suffered from the economic depression which reduced foreign trade drastically. Then there occurred the post-war reduction in North Sea herring stocks due primarily, it was believed, to the growth of industrial fishing with the purse seine in Norway. For these and other reasons, the majority of the fishing boats in Europe were engaged in demersal fishing in the early 1950's. Some boats used drift nets for a few months in the year, but drifting was a labour intensive fishing and was reckoned to be too risky a venture for all but the best fishermen.

At this time midwater trawling was very much in the experimental stage. Although Larsen had shown the way with the pair trawl, most research institutes were obsessed with single-boat midwater trawling, and much effort was expended before it became apparent that pair trawling was much more effective and more economical for all but the biggest of fishing vessels. Larsen's nets were made of cotton and this made them vulnerable to damage when catches were large, or when they were being shot in a heavy swell. It was not until nylon and terylene twines became available that the midwater trawl came into its own. The fine strong twines necessary to allow good water flow through the net had to be made of the best synthetic fibres.

For successful midwater fishing it was found as a general rule, that the bigger the mouth opening the better. And in this respect, the pair trawlers had a big advantage over the single-boat trawlers. It was possible to spread the net mouth horizontally using midwater otter boards, but the power required to spread them was considerable – nearly as much as to tow the net itself. This meant that a single-boat midwater trawler had to expend nearly half its towing power on the boards alone. The pair trawlers needed no otter boards and consequently most of their power was used to tow the net. Put very simply, this meant that a single-boat midwater trawler of 200 hp could pull a net suitable for only half that power, *ie* 100 hp. On the other hand, two 100 hp vessels could pull a net suitable for double their power, *ie* 200 hp. Thus, per engine hp relative to size of net, the pair trawl was twice as efficient as the single-boat trawl. This is an extremely crude calculation, but it illustrates the point. As the vast majority of fishing vessels had less than 200 hp, pair trawling was the obvious choice. Today it is reckoned that single-boat midwater trawling is not really effective for vessels with less than 500 hp, and is better suited for stern trawlers with over 1,000 hp.

There were other reasons why pair trawling was more efficient. Two vessels could search an area better than one, and as fish location is absolutely essential to midwater trawling this was a vital factor.

When towing the trawl, the two boats pass on either side of the fish school and not directly over it as a single trawler would. Herring are easily disturbed and tend to swim deep when vessels pass

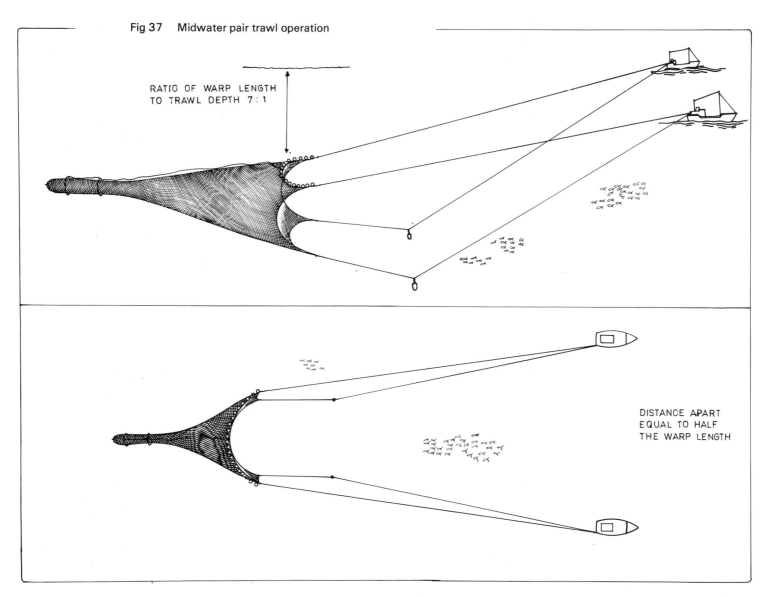

Fig 37   Midwater pair trawl operation

RATIO OF WARP LENGTH
TO TRAWL DEPTH 7:1

DISTANCE APART
EQUAL TO HALF
THE WARP LENGTH

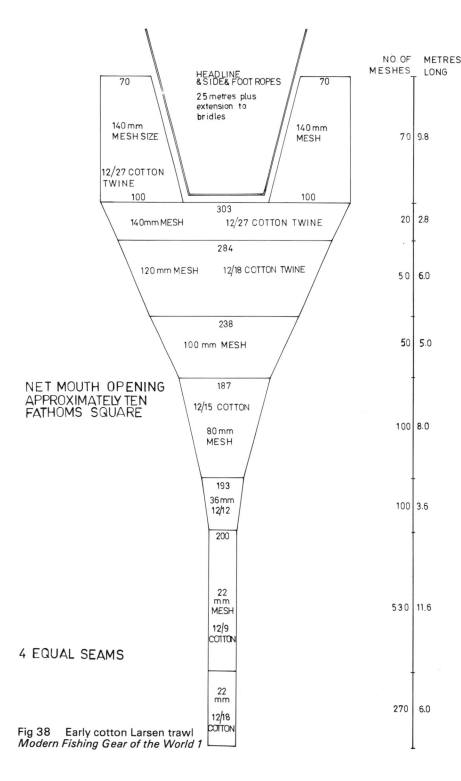

	NO. OF MESHES	METRES LONG
70 — HEADLINE & SIDE & FOOT ROPES — 70		
25 metres plus extension to bridles		
140 mm MESH SIZE — 140 mm MESH	70	9.8
12/27 COTTON TWINE		
100 — 100		
303 — 140mm MESH — 12/27 COTTON TWINE	20	2.8
284 — 120 mm MESH — 12/18 COTTON TWINE	50	6.0
238 — 100 mm MESH	50	5.0
NET MOUTH OPENING APPROXIMATELY TEN FATHOMS SQUARE — 187 — 12/15 COTTON — 80 mm MESH	100	8.0
193 — 36 mm — 12/12	100	3.6
200 — 22 mm MESH — 12/9 COTTON	530	11.6
4 EQUAL SEAMS — 22 mm — 12/18 COTTON	270	6.0

Fig 38 Early cotton Larsen trawl
Modern Fishing Gear of the World 1

over them, so in this respect also, pair fishing proved more suitable.

Common Misconceptions

Perhaps we can pause here to dispel some of the misconceptions about pair trawling. Although the method has proved to be a remarkably efficient and lucrative way of fishing for thousands of vessels in Europe, one still reads papers by learned and well meaning men who recite numerous 'disadvantages' to pair fishing that are simply not true. It is said that pair trawling is only suitable for daylight fishing because at night the vessels have difficulty maintaining the correct distance apart. This is just not true. The bulk of pair trawling operations are carried on at night, and have been so since the Larsen trawl was first used. A daylight tow is an unusual event for a pair trawler, unless it is for fish lying 60 fathoms deep or more.

It is said that because two vessels must share the proceeds from one net, pair fishing is not as economical as single-boat fishing. Well, what would the economist rather have – half of 20 tons or all of 5 tons? That is what it amounts to. Under the 500 hp limit, pair trawlers out-fish single vessels by such a margin that single-boat midwater trawling is deemed not worth the investment. Again it is said that to get two skippers to work together is extremely difficult. Now it is true that pair fishing demands close co-operation and team-work, but is this all that difficult to achieve? There are hundreds of skippers of independent mind working happily together with their partners, and some have been doing so for 20 years or more so the problems of a partnership operation are certainly not insurmountable. Another 'disadvantage' of pair trawling is that vessels must manoeuvre in close proximity when passing the bridles, and this is dangerous. It is dangerous, particularly in bad weather, yet the number of collisions, or even of minor bumps, is remarkably small. Most of the hundreds of pair trawlers in operation are covered by fully comprehensive insurance, yet to date no marine insurance company has even suggested higher rates for pair trawlers. This would indicate that in practice their operations are no more dangerous than those of other fishing boats.

Fishing Techniques

Midwater pair trawling techniques are essentially simple. But like all advanced methods of fishing, the rules must be correctly observed to ensure success. When only limited information was available, it took some pioneer skippers a year or more to master the techniques. It is relatively easy to use a new type of gear when there are dozens of vessels fishing successfully with it in the locality. It is not so easy for a pair of skippers who are starting from scratch and have no one nearby to whom they can turn for advice. Once the technique has been mastered, however, and the crews are familiar with the operation, then it is a fairly straightforward method of fishing. It is always exciting, however, as the net is never shot until fish schools are detected, and when fish are caught, it is usually in large quantities.

The first part of the operation is the search for fish

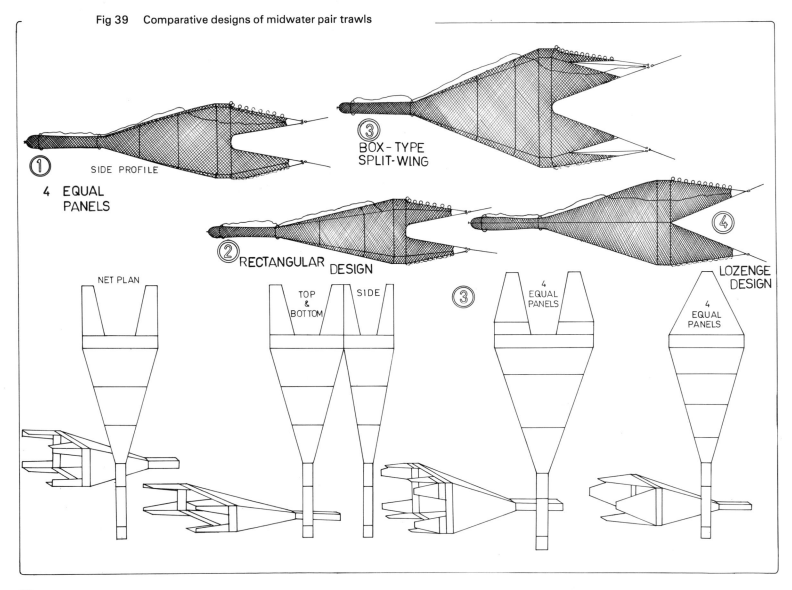

Fig 39   Comparative designs of midwater pair trawls

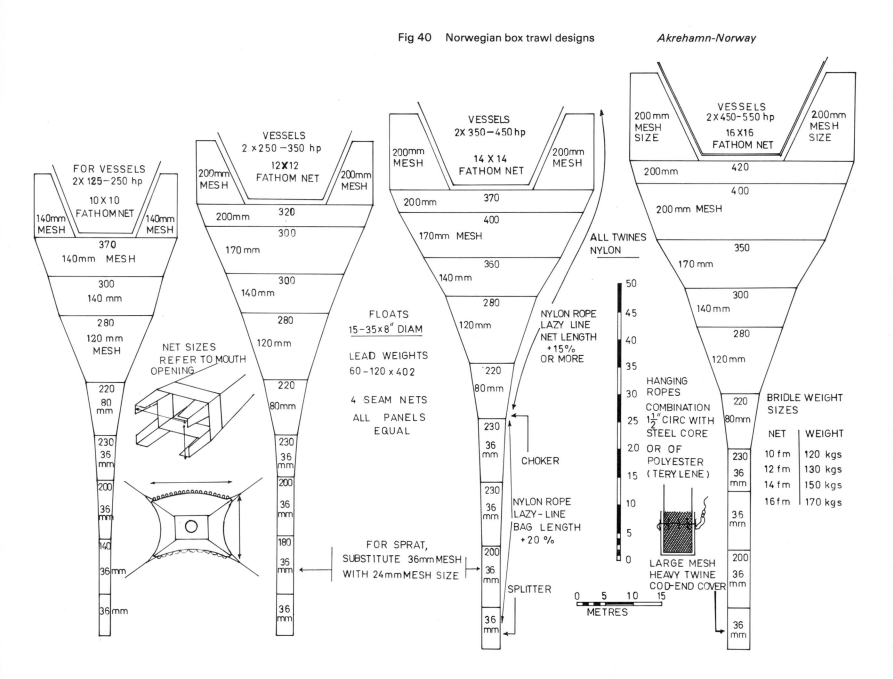

Fig 40   Norwegian box trawl designs   *Akrehamn-Norway*

FOR VESSELS
2X 125−250 hp

10 X 10
FATHOM NET

140mm MESH | 140mm MESH

370
140mm MESH

300
140 mm

280
120 mm MESH

220
80 mm

230
36 mm

200
36 mm

140
36 mm

36 mm

NET SIZES REFER TO MOUTH OPENING

VESSELS
2 × 250 −350 hp

12 X 12
FATHOM NET

200mm MESH | 200mm MESH

320
200mm

300

170 mm

300
140mm

280

120mm

220
80mm

230
36 mm

200
36 mm

180
36 mm

36 mm

FLOATS
15−35 x 8" DIAM

LEAD WEIGHTS
60 −120 x 402

4 SEAM NETS
ALL PANELS EQUAL

FOR SPRAT, SUBSTITUTE 36mm MESH WITH 24mm MESH SIZE

VESSELS
2X 350−450 hp

14 X 14
FATHOM NET

200mm MESH | 200mm MESH

370
200mm

400

170mm MESH

360
140mm

280
120mm

220
80mm

230
36 mm

230
36 mm

200
36 mm

36 mm

ALL TWINES NYLON

NYLON ROPE LAZY LINE NET LENGTH +15% OR MORE

CHOKER

NYLON ROPE LAZY-LINE BAG LENGTH +20 %

SPLITTER

VESSELS
2X 450−550 hp

16 X16
FATHOM NET

200 mm MESH SIZE | 200mm MESH SIZE

420
200mm

400

200 mm MESH

350
170 mm

300
140 mm

280
120mm

220
80mm

230
36 mm

36 mm

200
36 mm

36 mm

HANGING ROPES COMBINATION 1½" CIRC WITH STEEL CORE OR OF POLYESTER (TERYLENE)

LARGE MESH HEAVY TWINE COD-END COVER

BRIDLE WEIGHT SIZES

NET	WEIGHT
10 f m	120 kgs
12 f m	130 kgs
14 f m	150 kgs
16 f m	170 kgs

50
45
40
35
30
25
20
15
10
5
0

0   5   10   15
METRES

81

which usually begins in the late afternoon. The best time for herring schools is just after sunset and just before dawn. When one considers that the peak schooling periods may total only 4 or 6 hours in the 24, one will realise that the trawlers must seize the opportunity when it presents itself. If a pair of trawlers 'miss' a school by towing wide of the mark or at the wrong depth, they rarely get another chance of netting the same school. Midwater trawling is a hit or miss affair and one either gets a bag of fish or nothing at all. Sonar is coming more into use now but many pair fishers rely only on vertical acoustic sounders or fish finders. Once a school of fish is located, and its extent and depth determined, the vessels move downwind and one of them pays out the net. This is done very quickly. The other boat draws near and collects her bridle ends passed by heaving line. The bridles are connected to their

Fig 41   English, French and Scandinavian designs   *FAO Catalogue 1*

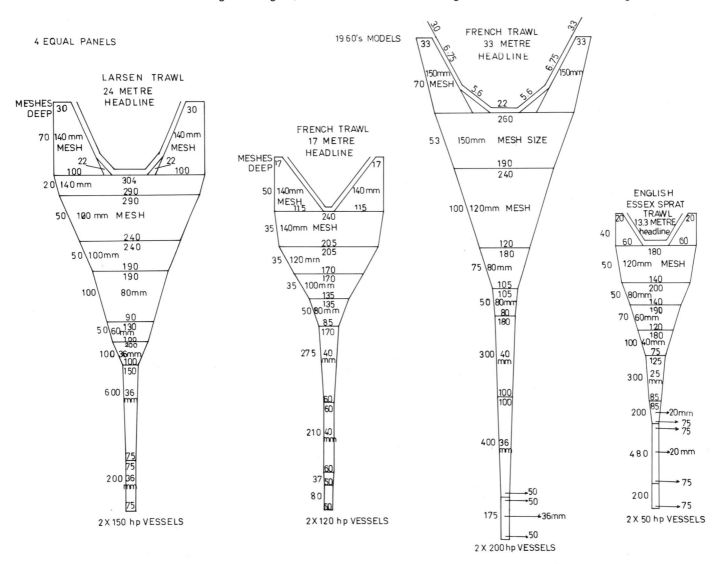

respective warp ends by means of quick-release hooks and the strain is taken on the winches. The lower warp weight is dropped a bit, the boats face the direction of the herring school, and at a given signal they shoot the warps at full speed. Speed is reduced as the last of the warp is going out, the winch brakes are made fast, and then towing commences with the vessels on parallel courses about half or a third of the warp length apart.

Towing speed is at or close to maximum continuous rpm.

It used to be thought that the chief difficulty in midwater trawling was controlling the depth of the net. This technique has been well mastered now, and the net depth is adjusted by increasing or decreasing the warp lengths according to a rule of thumb developed for the particular type of gear in use. Of course, as scientists will point out, depth can be

Fig 42   Rectangular mouth midwater trawl design

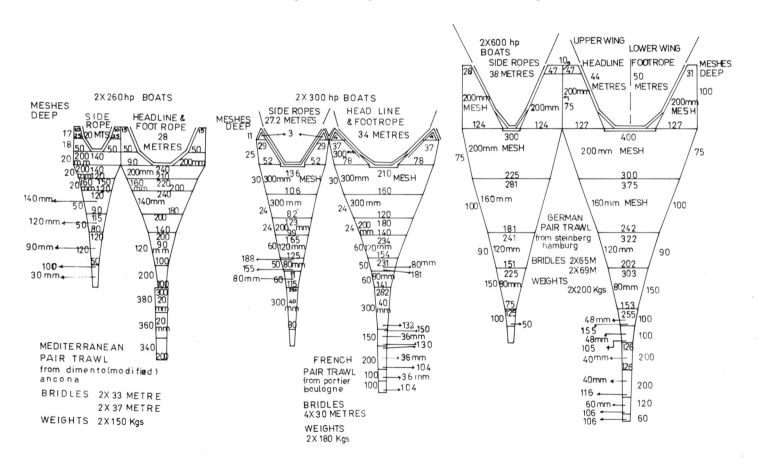

altered by increasing or decreasing speed, but as most pair trawlers tow at full speed, or near it, this is not really practical. Other methods of adjusting net depth include towing closer or further apart than the standard 'half-warp' length. The net can also be made to jump or rise over an obstacle by heaving in one or two fathoms of warp. This lifts it about 5 fathoms in 30 seconds. The same effect can be achieved by slacking the top wires only by the same amount. After the obstacle is cleared the lower wires are slacked out the same amount to square the gear.

The boats tow ahead while keeping a careful watch on their echo sounders. If one boat only marks fish, then the pair turn slowly in that direction. If both mark fish then they maintain their course. Once the net has been pulled through the school of fish, hauling begins. Tows generally last only about 15 minutes, rarely as long as an hour, and sometimes as short as 10 minutes. Once the warps and weights are up, the bridles are passed to the hauling vessel and the net is pulled in. A bag of herring may contain 10, 20 or 40 tons of fish and it

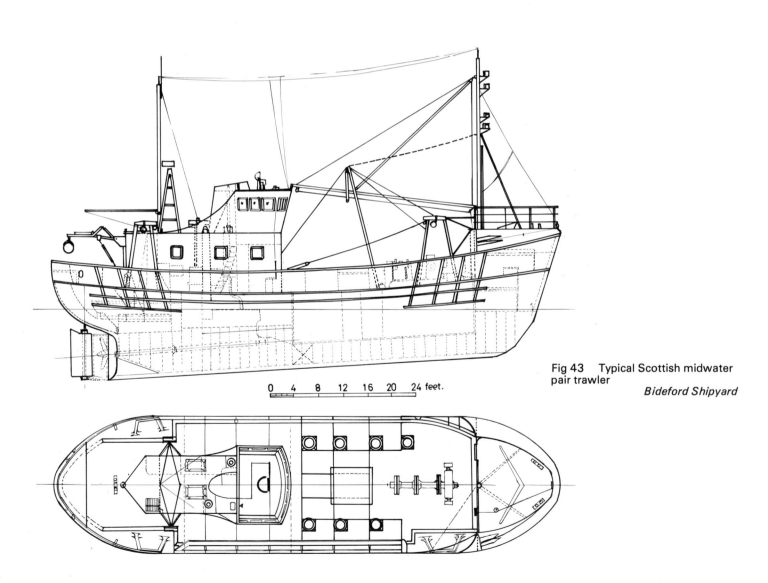

0 4 8 12 16 20 24 feet.

Fig 43   Typical Scottish midwater pair trawler

Bideford Shipyard

84

takes the hauling vessel some time to get these on board. This is where the 'group' system pays dividends. Vessels working in groups of 3, 4 or 5, can pair off with the 'free' vessel while the other is taking fish on board, or returning to the market.

Pair trawlers in Denmark, Britain and Ireland are mostly wooden vessels 60-85 feet in length, with 150 to 750 hp engines. The bulk of these vessels today are probably in the 75 foot 350 hp class. Larger pair trawlers operate from Germany and Sweden, some with as much as 900 hp but the total number of these

boats is small. Most pair trawlers are rigged like traditional side trawlers with the addition of a power block aft. Ordinary trawl winches are used but large capacity winches are better as considerable warp lengths are necessary for deep water towing. The wire warps used in midwater trawling are not as thick as those used for bottom trawling. Good manoeuverability is necessary, particularly for bigger vessels. For this reason some large side trawlers are not really suitable for pair fishing.

Fig 44   Gear and accessories used in midwater pair trawling

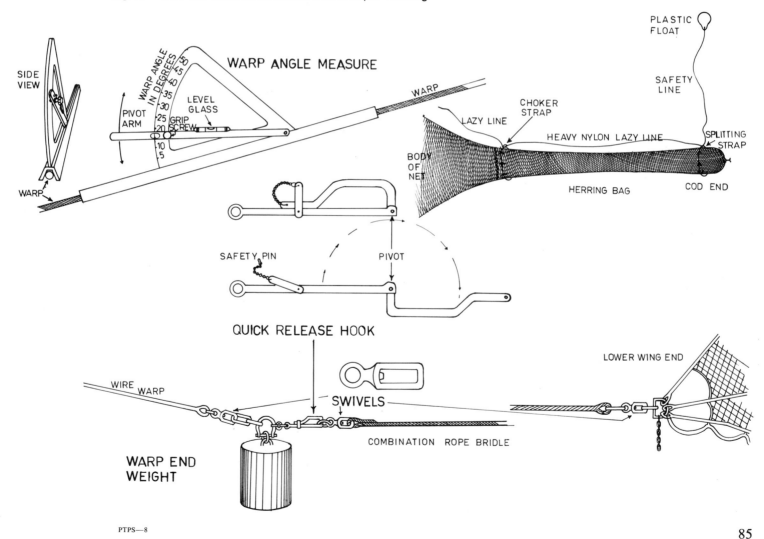

Midwater Trawl Nets

Midwater trawls are almost exclusively 'box' trawls having four seams instead of two as with most bottom trawls. The majority of nets are of Larsen's original type with four identical seams. Within this group there are many variations such as 'lozenge' types, split-wing types and so on. A large number of rectangular nets are also used. These have shorter side panels which result in a rectangular mouth opening instead of a square one. The rectangular net is particularly useful when towing in shallow water or when towing close to the bottom. Most manufacturers in Denmark and Sweden designate their net sizes by the estimated size of the square mouth opening in fathoms. For instance a 20 x 20 fathom net would be one with a mouth opening 20 fathoms square, *ie* 400 square fathoms, or 14,400 square feet. In Britain and some other countries the net size is given by the number and size of the meshes around the mouth. This could be 4 times 300 = 1,200

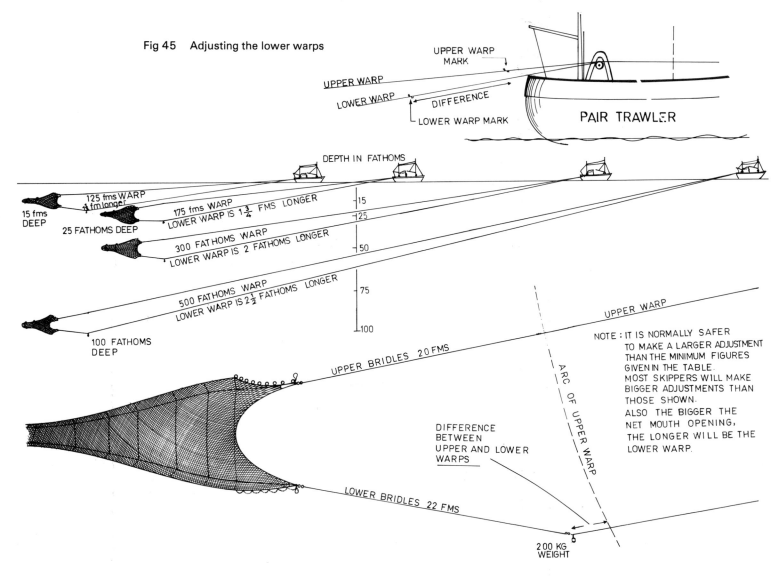

Fig 45   Adjusting the lower warps

UPPER WARP MARK

UPPER WARP

LOWER WARP

DIFFERENCE

LOWER WARP MARK

PAIR TRAWLER

DEPTH IN FATHOMS

125 fms WARP
1 fm longer
15 fms DEEP

175 fms WARP
LOWER WARP IS 1¾ FMS LONGER
25 FATHOMS DEEP

300 FATHOMS WARP
LOWER WARP IS 2 FATHOMS LONGER

500 FATHOMS WARP
LOWER WARP IS 2½ FATHOMS LONGER

100 FATHOMS DEEP

15
25
50
75
100

UPPER WARP

UPPER BRIDLES 20 FMS

ARC OF UPPER WARP

NOTE : IT IS NORMALLY SAFER TO MAKE A LARGER ADJUSTMENT THAN THE MINIMUM FIGURES GIVEN IN THE TABLE. MOST SKIPPERS WILL MAKE BIGGER ADJUSTMENTS THAN THOSE SHOWN. ALSO THE BIGGER THE NET MOUTH OPENING, THE LONGER WILL BE THE LOWER WARP.

DIFFERENCE BETWEEN UPPER AND LOWER WARPS

LOWER BRIDLES 22 FMS

200 KG WEIGHT

Fig 46   Factors relating to warp adjustments

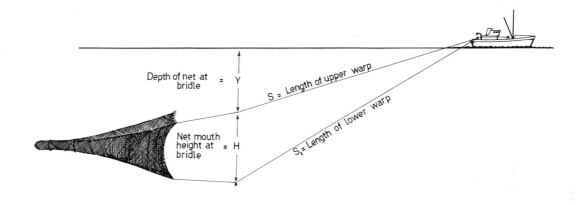

Depth of net at bridle = Y

S = Length of upper warp

Net mouth height at bridle = H

S₁= Length of lower warp

Fig 47   Warp adjustment for 15-fathom trawl

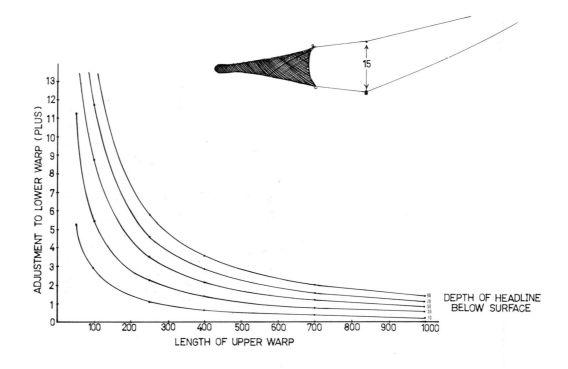

15

DEPTH OF HEADLINE BELOW SURFACE

ADJUSTMENT TO LOWER WARP (PLUS)

LENGTH OF UPPER WARP

meshes 200 mm stretched mesh (8 inches). To estimate the mouth opening from this, one usually takes half the mesh size, so: 8 inches ÷ 2 = 4 x 300 = 1,200 inches or 100 feet, per side, squared equals 10,000 sq. feet or about 280 sq. fathoms.

The first midwater herring trawls had rather small meshes in the wings and around the mouth, 140 mm or 5½ inches. Recently it was discovered that the herring could still be guided into the net mouth by extremely large mesh wings providing they were tightly hung. Mesh sizes jumped to 300 mm (12 inch), 600 mm, and even one metre or more. Some midwater trawls are fishing well with 6 foot (1.83 metre) meshes in the wings. This large mesh netting is of course made with heavy nylon twines, sometimes braided nylon line. By using such large meshes, pair trawlers are able to increase substantially the size of their net mouth without increasing the towing load. Sometimes a panel of large meshes is inserted half-way down the bag to assist the water flow through the net. One such net by Robert Larsen, designed for 350 hp vessels, has a mouth opening of 24 x 24 fathoms, compared to one of 19 x 19 fathoms for the same size of vessel, but

using smaller meshes. The mouth area of the big mesh net is 570 sq. fathoms compared to 360 sq. fathoms for the smaller mesh net.

Midwater trawls are hung to their ropes at similar ratios to high-opening bottom trawls. The 'square' is mounted with about 50% slack or more but the wings are mounted very tight with practically no slack in the webbing. If a bag line is mounted down the seam selvedge for strength, then it is made equal to the stretched length of netting or very slightly less.

One feature common to most midwater trawls is that the top sheet or panel of netting is kept slightly tighter than the lower sheet. This is achieved first by reducing the top panels by one row (half a mesh) each when the net is assembled. The net is checked regularly by the crew who stretch out the upper and lower panels on the pier and measure them against each other. If the top sheets have stretched (which they do) longer than the lower sheets, then they are shortened appropriately. Sometimes it is easier simply to turn the net 'upside down' and change the floats with the footrope chains. It is important to check the hanging of midwater trawls regularly because they will not catch fish if they are out of

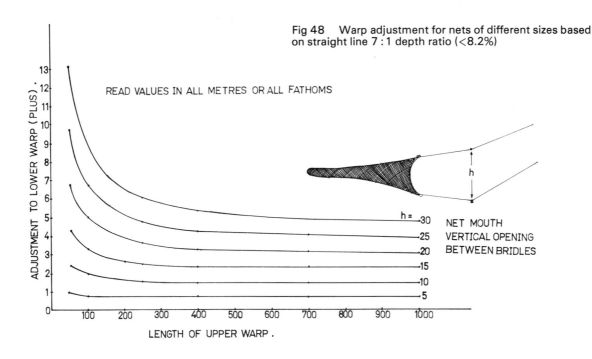

Fig 48   Warp adjustment for nets of different sizes based on straight line 7 : 1 depth ratio (<8.2%)

trim, and with bags of fish weighing up to 100 tons, it is to be expected that the netting will stretch. In addition to checking the net it is important that the bridles are measured regularly to make sure that they are even.

As regards twine material, there are two schools of thought relating to the net design. The Scandinavian fishermen who pioneered the gear preferred to have a fine, strong twine that would not stretch. For this they selected polyester (terylene) which possesses these qualities better than any other fishing net twine. Sometimes low-stretch nylon was used. This is an ordinary nylon twine constructed in such a way that it will have minimum elongation. The Scandinavians claim that nets made of such material can be trimmed more accurately and will fish better. This is probably true, but the fine twines make them vulnerable to damage, and when one of these nets tears or 'explodes' it suffers extensive

damage necessitating long major repairs.

The second school of thought, held by Dutch and German fishermen, is that a strong stretching twine is best to withstand the tremendous strains of midwater fishing and that this more than compensates for any lack of precision in the trim and rigging of the gear. They use a special nylon twine that is constructed with good elongation properties. The twine is usually thicker and stronger than that used in Scandinavian trawls. Skippers using such gear claim that it is less liable to be damaged or to burst with the weight of fish or the force of water when shooting in a heavy swell. Generally speaking the bigger, more powerful pair trawlers use the Dutch-German type gear and the smaller, lighter powered craft use Scandinavian type nets. (This does not mean Scandinavian gear is not suitable for big vessels. There are Swedish pair trawlers with 1,000 hp or more in each vessel). The Dutch

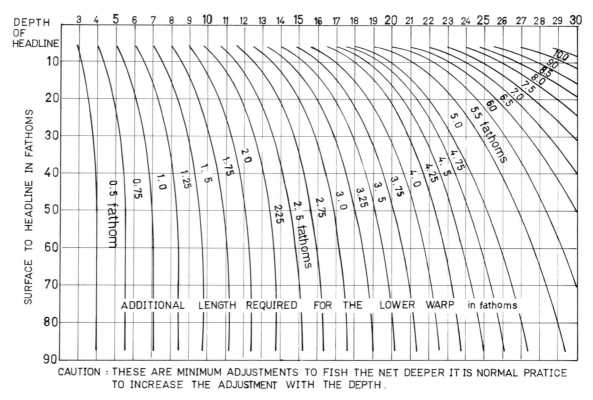

Fig 49   Warp length adjustment graph

and German boats use more of the 'rectangular' mouthed trawls with smaller side seams.

It is now generally agreed that nets of high elasticity are better to withstand the stresses created by large catches, and by the generally increased towing power of the average pair of trawlers. The German and Dutch twines developed for this

purpose were adopted to such an extent in Britain that the manufacturers there had to begin to produce such material themselves. This they now do with good effect. Braided nylon ropes are also used as hanging lines because of their better elasticity.

The net designs were further modified to reduce damage when towing close to the sea bed. Previously

Fig 50   Midwater pair gear for large trawlers
J Scharfe NFLD 1966

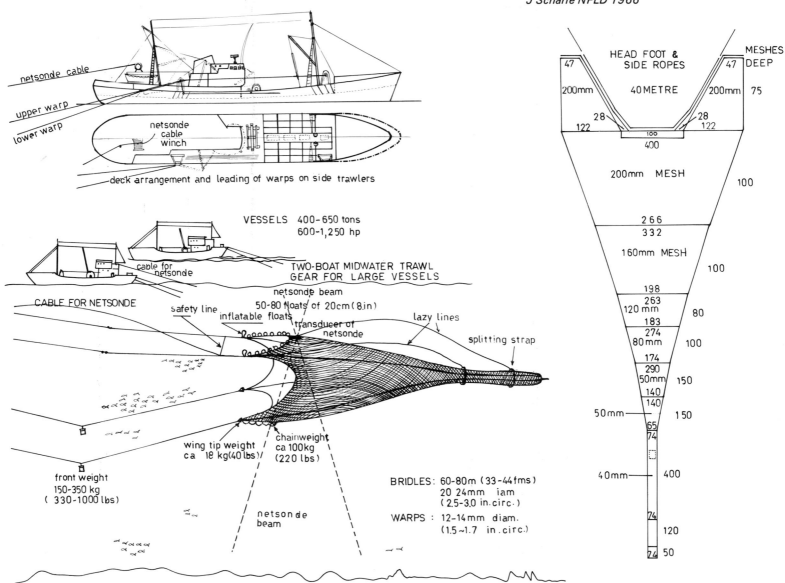

a taper cut of 1 point 2 bars was used all the way from the wings to the narrow end of the bag but in practice this appeared to result in excessive sagging in the bellies. To eliminate this, manufacturers began using a cut of all points in the wings, followed by a steep cut of 1 point 4 bars in the body. This appeared to result in a much better net formation with a minimum of loose webbing in the belly, and consequently less danger of fouling on the sea bed.

To achieve vertical mouth opening, floats and weights are used. But the floats and footrope weights actually contribute only a small part of the opening force. The vertical opening is really achieved by the upward pull of the top warps and the downward pull of the lower warp-end weights. These weights may be anything from 100 to 350 kilograms each depending on the power of the

vessels and the size of the net. With the early pair trawls, the weights were attached at the wing-ends of the net, but later it was found to be better to place them between the bridle and the warp. On some small pair trawls, particularly sprat trawls used in shallow water, the weights may still be attached at the net but with nearly all herring trawls they are on the end of the warp, 15–30 fathoms or more from the net. Note that the weights are shackled to the lower warp end, not to the bridle. The bridle is connected to the weight by means of a quick-release hook which permits the rapid disconnection of the bridles when they are passed to the hauling vessel. The weights may be made of iron, lead, concrete or chain. Sometimes smaller weights of 10 or 20 kilos are attached at the wing ends, and this, plus a light chain or leaded rope, completes the net weighting.

Fig 51   Canadian two-boat capelin trawl
G Brothers 1974 IDB Newfoundland

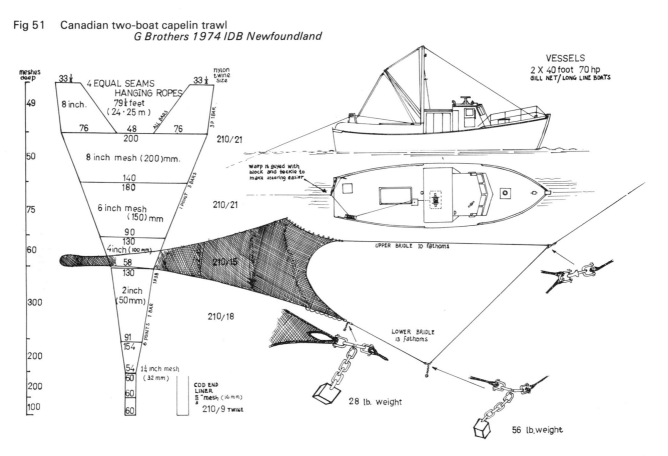

Lower Warp Adjustment

The lower pair of bridles are usually from 2 to 4 fathoms longer than the upper pair. This is to compensate for the distortion of the lower warp and bridle which will have to be longer than the upper ones. Sometimes equal bridles are used but then the lower warps must be lengthened by a greater margin after shooting the gear. The amount of extra warp paid out will depend on the depth to which the net is set. The deeper the water, the longer the lower warp must be relative to the upper warp. Each pair of vessels will work to a rule of thumb based on a calculation for the type of gear and length of bridles they are working. With a set of lower bridles 2 fathoms longer than the upper ones, a pair of vessels may use 2 more fathoms of lower warp at 30 fathoms depth, 1 more at 20 fathoms and equal warps at 10 fathoms. If the fish schools are less than 10 fathoms from the surface they may make a 'negative'

adjustment with the upper warps slightly longer than the lower ones. The precise adjustment to be made will depend on the size of net used and whether the bridles are of equal lengths or not. It can be calculated trigonometrically providing one knows the distance between the upper and lower bridles and the depth of each bridle from the surface. The formula could then be:

$$\text{Lower warp length} = \sqrt{Dl^2 + W^2 - Du^2}$$

in which Dl is the depth of the lower bridle, Du is the depth of the upper bridle and W is the length of the upper warp. The depth of the bridles can be calculated by the warp length and the angle of the warps at the vessel. There is a simple device used to measure the warp angle at the boat. It consists of a large brass protractor with a glass 'level' at its straight side, and a pivoted arm which can be screwed in place when it is placed along the warp with the level glass reading horizontal. Needless to

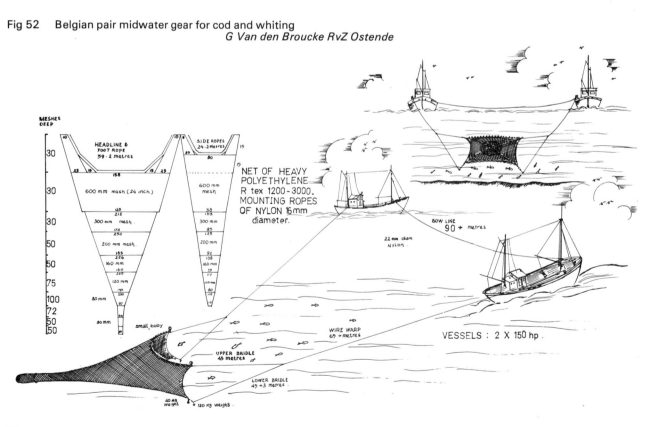

Fig 52 Belgian pair midwater gear for cod and whiting
G Van den Broucke RvZ Ostende

say, accurate readings can only be obtained in fine weather and even then an allowance must be made for warp distortion.

In actual practice warp adjustments are arrived at empirically. When a new set of gear is to be fished, the lower warp lengths are estimated trigonometrically. After some practice, the skipper later makes up his own adjustment table (a mental one only). From countless observations on the gear performance he can tell whether the gear fishes better with 2½ fathoms extra on the lower warp, or with more, or less. In any case, the warp adjustment need not be precise. A few feet either way will not make a great deal of difference.

Warp Length Ratio

More important to the skippers is the matter of relating warp length to depth. As most pair trawlers tow at, or close to, full speed, there is little room for depth adjustment by speed change. In any case, too slow or too fast a towing speed could adversely affect the fish-catching power of the net, so warp length is usually the means used to obtain net depth. Here again there is no general rule as what works for a pair of 150 hp boats will not necessarily work for a pair of 950 hp boats. However, a ratio of 7 : 1 is commonly used, beginning at about 10 fathoms depth because pair trawlers never set less than 75 fathoms of warp. The length of the bridles is not included in the warp length, and if there is any doubt about the depth, skippers are always inclined to release more warp rather than less. This is good practice with herring as they tend to swim deeper when disturbed. A lot will depend on the power of the vessels and the size and weight of the gear, but the 7 to 1 ratio is probably the most common.

In deeper water, pair trawlers may use a lower warp to depth ratio, sometimes as low as 4 to 1 though that is the extreme. This is due more to the lack of warp than anything else. Most trawlers carry only about 350 or 400 fathoms of wire on each trawl drum, and this would enable them to fish as deep as 50 or 60 fathoms. However, if fish are found at 90 fathoms for instance, it may be possible to reach them with only 400 fathoms of wire, providing they use heavier weights or (more likely) tow at a slower speed. Net depth can also be increased by towing more closely together, ie at a quarter or a third of the warp length apart, rather than a half which is most common. With experience, a pair of skippers can usually place their trawl accurately. A net transducer would help, of course, but once the skippers are used to the gear it would not be really necessary. Vessels which use them regularly do so to tell them whether fish are entering the net mouth, as well as to determine the net depth.

Midwater pair trawling can be harder on an engine than bottom trawling because speed is essential and the huge net is heavy to tow. Vessels therefore tow the gear with the engine at maximum continuous rpm. It is better for the engine if towing could be carried out at 75% of maximum engine power. This would leave the vessels with a burst of extra speed to be used when schools are ahead of the net. For this reason, some skippers when building a new boat for pair trawling, install an engine with 25% more power than they intended to have. This reserve power is very handy when fishing and it helps to prolong engine life. It might be appropriate to mention the balancing of the power of pair trawlers at this point. Many people are under the impression that the boats must have equal power in order to fish successfully. This is not true. Many successful 'pairs' have different engine sizes, sometimes one vessel may have twice as much power as the other. Both vessels tow the gear from the gallows or tow-block on the after starboard side. This is for good reasons of manoeuverability. As a result of towing on the same side, one boat has the 'heavy' end of the gear. This is the 'starboard' vessel whose warps lead across the stern. It must use more helm than the port vessel, and consequently it uses more power. As a result of this, if the two boats do not have the same size of engine, the more powerful vessel always takes the starboard side of the gear.

Table of Warp Angles and Estimated Net Depth

Warp Angle	100	150	200	250	300	325	350	375	400
4°	7	10½	14	17½	21	23	25	27	28
5°	8¾	13	17½	21½	26	28	31	33	35
6°	10½	15¾	21	26	31	33	36	39	42
7°	12	18	24	30	36	39	42	45	48
8°	14	21	28	35	42	45	48	51	55
9°	15½	23	31	38	46	50	53	57	60
10°	17	25½	34	42	51	55	59	63	66
11°	19	27	38	47	57	61	65	69	73
12°	21	31	42	52	63	66	70	75	80
13°	22½	33	45	55	66	70	75	80	85
14°	24	36	48	59	70	75	80	85	90
15°	26	39	52	64	75	80	85	90	97
16°	28	41	54	67	80	85	90	95	100

(Warp length in fathoms. Depth of net in fathoms (approximate))

The net depth will be reduced proportionately in deep water, due to the curvature of the warps

Notes on Lower Warp Adjustment

It is not possible to give fixed values for the lower warp adjustments because of the differences in net design and rigging. When purchasing a midwater trawl one should ask the netmakers' advice on adjustments for that particular model.

Warp end weights differ according to the type of net and the power of the vessels involved. The amount of weight, the length of bridles, and the size of the net, will all contribute to the form of the gear in operation; and this will be further affected by the towing speed and the distance the vessels maintain apart.

A fair guide to weight relative to net size, is given in the following table by Leslie Innes:

Vessel hp (each)	Net size in fathoms	Warp end weights (each)
120-250	20 x 20	165-220 lb. (75-100 kilos)
250-350	25 x 25	220-275 lb. (100-125 kilos)
350-450	30 x 30	275-330 lb. (125-150 kilos)
450-550	35 x 35	330-385 lb. (150-175 kilos)

The lower warp adjustment must be reckoned against the difference in length in the upper and lower bridles. The lower bridles may be 2, 3, 4 or even 5 fathoms longer than the upper set. Some boats operate with bridles of equal length. Caution must be exercised when adopting a set of adjustment figures to see whether they apply to unequal bridles, and if so, by what margin. Naturally, with a larger-mouth net, the adjustments will have to be larger. Also, when fishing in deep water with limited warp length, it is possible to make the net sink down by increasing the lower warp adjustment. In most cases, skippers prefer to give more than the tabulated allowance.

The Skagen company of Christensens gives the following advice for adjustments when using their 34 x 24 fathom trawls.

Firstly, the lower bridles are made 4 fathoms longer than the upper ones. Then:

Warp length in fathoms	Adjustment in fathoms or (metres)
50	top wire + 2 fms. (4 metres)
65	top wire + 1¾ fms. (3 metres)
75	top wire + 1½ fms. (2·7 metres)
87	top wire + 1¼ fms. (2·2 metres)
100	top wire + 1 fm. (1·8 metres)
125	top wire + ¾ fm. (1·4 metres)
150	top wire + ¼ fm. (0·45 metre)
175	both wires equal
200	lower wire + ¼ fm. (0·45 metre)
225	lower wire + ½ fm. (0·9 metre)

Using short wires, the warp end weights are 750/875 lb. (300/350 kilos)
Using long wires, the warp end weights are 1,000/1,125 lb. (400/450 kilos)

With fish located 6 fathoms deep, they use 50 fathoms warp
With fish located 10 fathoms deep, they use 100 fathoms warp
With fish located 14 fathoms deep, they use 150 fathoms warp
With fish located 30 fathoms deep, they use 225 fathoms warp

This data was derived empirically by fishermen, and is given only as a guide.

94

Plate 18   Midwater pair trawlers in New Zealand pass the heaving line for the bridles   *Australian Fisheries*

Plate 19   Two vessels join in hauling the net   *Australian Fisheries*

Plate 20   Taking the catch aboard
Australian Fisheries

Plate 21   Final stages of hauling in the catch
Australian Fisheries

Sample Table of Lower Warp Adjustments for a Net with a Ten Fathoms Opening between the Upper and Lower Bridle Ends

Depth of net (upper bridle)	Warp length in fathoms (upper warp)								
	100	150	200	250	300	350	400	450	500
5	1	½	¼						
10	1½	1	½	¼	¼				
15	2	1⅓	1	½	½	¼			
20	2½	1⅔	1¼	1	¾	½	¼		
25	3	2	1½	1¼	1	¾	½	¼	¼
30		2⅓	1¾	1½	1¼	1	¾	½	½
35		2⅔	2	1½	1⅓	1¼	1	¾	⅔
40		3	2¼	1¾	1½	1⅓	1¼	1	¾
45			2½	2	1⅔	1½	1⅓	1¼	1
50			2½	2¼	1¾	1⅔	1½	1⅓	1¼
55			2¾	2⅓	2	1¾	1⅔	1½	1⅓
60			3	2½	2¼	2	1¾	1⅔	1½
70				3	2½	2¼	2	1¾	1⅔
80					3	2½	2¼	2	1¾
90							2⅓	2¼	2
100								2⅓	2¼

Additional Length of Lower Warp in fathoms (minimum adjustments)

Use of Headline Transducers

The control of the fishing depth of the net is a complex and difficult problem. Since the distance between the vessels, their speed, the warp length, and the warp adjustment, are all important factors, control cannot be achieved by attention to just one of the variables. Experience gathered empirically is most useful, but it takes time and the results usually relate to only one type of net, size of vessel or technique. The major problem is the precise depth of the net is never certain. But with modern instruments this problem can be overcome.

Headline transducers attached to the mouth of the net, and connected by cable to a recording display in the wheelhouse can provide the data needed. The transducer can be arranged to provide depth readings from the bottom or from the surface. They can also give an accurate indication of the vertical mouth opening and will show fish entering the net (or missing it, if the depth is wrong). As the skipper is

Plate 22   Model for four seam midwater trawl
Iver Christensens

able to tell the depth of the schools from his other echo sounder, he can place the trawl at the precise depth required to catch them. The value of the instrument is therefore enormous. The best modern pair fishers use the net sounder constantly.

There are drawbacks, however, as greater care is required when shooting the gear, and when towing and hauling, to ensure that no excessive strain is placed on the cable, also that it is not too slack, for then it could foul the net. But the techniques of using headline transducers are not too difficult and most of the successful pair fishers now have this equipment.

Some cable-less transducers are now available and no doubt as the technological problems are resolved, these will replace the cabled types. Where cables are used, a special cable winch is fitted to accommodate the wire. This winch is less powerful than a trawl winch, some even operated by hand, and they have brakes that can be set to release more cable if the tension rises too high.

Midwater Pair Trawling Operation

It takes some experience to manoeuvre a pair of vessels when towing. If the boats are to make a sharp turn, then the 'inside' vessel must slow down while the 'outside' boat maintains full speed. As they both tow on the starboard side, it is easier to turn to starboard than to port. The best way to make a 180 degree turn is for the vessels to haul up all but 25 or 50 fathoms of warp, then to complete the turn and shoot the warps again after the boats are on the new towing course. It is essential that the skippers maintain radio contact all the time. As the normal radio-telephone channels are already overcrowded, vhf radio is mostly used during the trawling operations. Some skippers use 'walkie-talkie' radios which have sufficient range for pair trawling (half a mile). They exchange information on warp length, lower warp adjustment, towing course, engine speed, fish marks and other vessels' activities. The skipper of the 'shooting' vessel, or the 'commodore' skipper, will make the final decisions regarding tactics and hauling time. Once a pair of skippers get to 'know' each other and develop team-work, then the operation becomes fairly routine with a

minimum of orders necessary.

Most pair trawlers operate as single pairs, with no other partners. They share the proceeds from the catch evenly, and then each vessel deducts its operating expenses and pays its crew, from its own share. It may happen that one boat may be 'free' when its partner is taking a big bag of fish on board, and it may then pair up with the free vessel from another pair in a similar situation, for a single tow. In such a case, the proceeds from that tow are shared out on an *ad hoc* arrangement determined by the skippers themselves. Some pair trawlers operate in groups of three, four or five. In such instances there is a more careful sharing of investment with each skipper purchasing trawls and equipment of a design and to a value agreed to by the group. The gross earnings of the whole group are shared equally by the vessels which then pay for their own operating expenses unless there is a large discrepancy in fuel consumption, which may necessitate all fuel costs being equally borne by the group members. The 'big five' arrangement as it is sometimes called, requires considerable team-work to succeed, but when properly carried out, its advantages are enormous. Fishing time per vessel is increased, steaming, landing and turn-round (refuelling, *etc*) time is kept at a minimum by loading up the boats in turn. The five vessels can search an area and locate fish much more quickly than a single pair. Some such groups of boats engage in herring pair trawling nearly year-round, following the fish migrations around the coast.

Pair Trawling with Larger Vessels

Midwater pair trawling by large vessels is conducted chiefly from Sweden and Germany by trawlers of 400–650 tons, 600–1,250 hp. The vessels must have good manoeuvrability for this work, and not all side trawlers are really suitable. The more powerful boats can tow a net with a 20 metre square mouth opening, using 200 mm meshes. With bigger meshes, a much larger net mouth can be obtained. The large vessel pair gear differs from that used by small boats in that heavier twines and ropes are used, but basically the gear design and operation is the same. German trawlers use a safety wire 25 mm diameter across the net mouth to prevent the net from taking excessive tension if the boats move too far apart. Many large trawlers are equipped with net transducers and cables (Netsonde) which are plugged into echo sounders on the bridge to give a constant picture of the net, its mouth opening and depth, and the fish schools (hopefully) entering the net mouth. Net transducer cables are a bit awkward to shoot from side trawlers but the German skippers have mastered the techniques satisfactorily. With their extra power and length of warp, such large pair trawlers have been able to fish successfully at depths up to 200 fathoms. Bottom trawlers of that size would normally use warps of 20–25 mm diam. ($\frac{3}{4}$ inch–$\frac{7}{8}$ inch) but for midwater trawling, thinner warps are preferred, usually 16 mm diam. ($\frac{5}{8}$ inch), which are easier to tow. The big pair trawlers use more of the rectangular-mouth designed nets than the traditional trawl with four equal seams, but both types are in use.

Fish Behaviour

Most pair trawlers operate at night on active herring schools within 40 fathoms of the surface. Spawning herring are much easier to catch but for most of the year the fish are active and will swim quickly to avoid the net. Measurements made with acoustical equipment and underwater cameras indicate that the herring can swim fast over short distances but it is doubtful whether they can maintain good speeds for any prolonged period. Midwater trawlers tow at speeds of 3–4 knots. In some cases where a fleet of vessels is operating in the same area, it has been found that the most powerful pairs catching the highest proportion of large good quality herring.

Daylight fishing for deep water schools requires more experience and skill than the near-surface trawling. Herring may be found in deeper waters, sometimes near the sea bed in daylight hours. In such instances, the schools are very scattered and may be difficult to detect by echo-sounder, except as 'smoky' marks. For these fish, pair trawl skippers make longer tows, perhaps 2 or 3 hours at a time. If the boats do not have sufficient warp to maintain a high warp/depth ratio, then they may tow at a

slower speed. Great care must be exercised if the fish are lying close to the bottom, particularly if the ground is rough or hard. It is possible to tell from the vibration on the warps whether the weights have touched the sea-bed and this would give some warning to vessels not equipped with net transducers. Once the techniques of deep water fishing have been mastered, pair trawlers can operate 'round the clock' so to speak instead of only at night during the near-surface schooling period.

Many people ask whether midwater trawls can catch fish species other than herring, sardine and sprat. Demersal fish sometimes form dense schools well off the sea-bed when preparing for migration or spawning. In such instances they can be caught by midwater trawl, but in temperate waters this happens rather seldom. When it does, however, species like haddock, whiting and coalfish (saithe) can be caught in large quantities. However, there is no sure way of telling whether a school of fish seen on an echo-sounder is composed of pelagic or demersal species, other than by catching them. Japanese vessels have been able to catch shrimp and squid in midwater trawls, and purse seiners have taken large quantities of snapper fish at times, so it is possible that midwater trawling in tropical or semi-tropical waters could be successful on semi-pelagic species. Around the British Isles, however, it is not common for demersal fish to be taken in a midwater trawl though occasionally species like mullet, saithe, haddock and whiting may be caught.

Multi-Purpose Vessels

Because of the seasonal nature of the herring fisheries, most pair trawlers are multi-purpose vessels which can convert to bottom trawling, Danish seining, ring netting or even purse seining during the 'off season' for midwater trawling. As there is very little difference in deck layout between a side trawler and a pair trawler, only a change of fishing gear is required to convert the vessel. For Danish or Scottish seining and ring netting, the after gallows is usually removed. Power blocks are applicable to all these methods of fishing and can be applied to each with good effect.

Some midwater pair trawlers carry large crews,

Plate 23   A bag of herring just surfacing   *D M Baikie*

perhaps eight or nine men as opposed to five or six for a bottom trawler of similar size. The extra men are required to handle the fish rather than the gear as such. Good quality fresh herring can command excellent prices on the market provided they are well cared for. To maintain high quality, pair trawlers box their herring (or as much of the catch as possible) in aluminium containers holding around 25 or 30 kilos each, with ice. Herring are also landed from bulk storage in fish pounds or lockers, but these fish tend to suffer from the pressure or weight of fish, and the absence of ice. In recent years there has been a lot of experimentation with chilled sea-water containers which can keep the herring in first class condition. The cheapest and most convenient system appears to be that using simple ice-cooled

tanks from which the fish are 'brailed' or removed with a large dip net. Some vessels, however, have successfully used removable containers which are transported by truck direct to the merchant who may be based in another country.

World-wide Use

Midwater pair trawling is practised today chiefly in Europe, but also in Canada, the eastern United States, and Australasia. Along with the purse seine, the pair trawl has displaced the drift net as the chief means of capturing herring. Unlike the purse seine which is an instrument primarily used in industrial fishing, practically all of the pair trawl catch is for human consumption.

Sweden is reckoned by most fishermen to be the real 'home' of midwater pair trawling. The Swedish fishing fleet is not large compared with that of neighbouring countries, but in terms of skill and technical advances, it is a world leader. The Swedes and the Danes pioneered bottom seining, wing trawling, and two-boat midwater trawling. Their modern pair trawlers are generally more powerful than their Scottish or Irish equivalents. One of the top vessels in the Swedish fleet was the *Tor-On* GG 204, a steel trawler over 33 metres long, 170 tons, and powered by a Nohab Polar 900 hp engine. The *Tor-On* can carry 180 tons of bulk fish or 1,500 forty-kilo boxes, in her two fish holds. Along with her partner vessel *Ganthi,* she worked a large midwater trawl net with 20 foot size meshes in the wings. Skipper Ahlstrom and the owners bought two new 38 x 8.2 metre vessels with 2,000 hp engines in 1976. This indicates the potential of midwater trawling with a pair of large vessels providing they are designed for the job. A pair of large British side trawlers attempted pair fishing for herring in 1973-75 but although they did enjoy some success, the vessels proved to be less than ideal for the job.

Fishing Report: Midwater Pair Trawl 1975 West Scotland

Most pair trawlers work in teams. One of the Scottish teams has a purse seiner as a team partner — the *Aquila* FR 105 owned and skippered by Jim Slater of Rosehearty. One of her pair trawl partners is the *Aquarius* FR 55, skippered by Peter Duthie from the same village near Fraserburgh.

The *Aquila* is powered by a 850 hp Caterpillar engine and is equipped for pursing with a net made by Stewart & Jack of Edinburgh – its dimensions are 90 fathoms deep and 400 fathoms long. *Aquarius,* with a 500 hp Stork-Werkspoor engine, is rigged for trawling with two nets 'laid on' ready to shoot on her wide transom stern. One net is 60 fathoms square with 64 inch mesh in the wings for towing in 'spreckles' (scattered shoals) in deep water: the other is 40 fathoms square, with 32 inch mesh in the wings. They are used with 120 lb. toe weights and top sweeps of 2 inch Belfast combination rope 10 fathoms long, with bottom ones of 15 fathoms; the difference in length ensures the net mouth is at right angles to the sea-bed when towing.

The warps which the two boats carry are 1¾ inch wire 500 fathoms long and marked every 25 fathoms. The weight on the bottom warp is 600 lb. made up of six lengths of heavy anchor cable, each weighing 100 lb. Jim and Peter have found that scrap anchor cable is easy to handle, does very little damage to the boat's side, and does not snag on rough, hard sea-bed.

They also carry two similar nets as spares, in pounds on either side of the wheelhouse.

Peter remembers the first trip of 1975 . . . It was cold and wintry. The boats had been moored in the basin at the Crinan Canal while the fishermen spent their New Year break at home in Aberdeenshire. It was now 6th January.

We left the canal at 10.30 a.m., the wind was force 4, so we decided to make for the Cruachan Bank, off Barra; it is usually a good bank for herring at that time of the year, and with a north west wind, we would have a shelter for working. There had also been reports of good shoals in the area. We took eight hours to do the journey down the Sound of Mull, up past Coll, and across the South Minch.

We arrived at half past eight in the evening. Listening to the vhf as we neared the grounds, we heard that boats already there had had good hauls at 'the close' (dusk), but by the time we arrived there was nothing and about 50 boats were steaming about. So we decided to search down along the shore towards Barra Head. At 10 p.m. our sonar (Simrad SK2) located a good shoal 3¼ miles off Mingalay.

We like to tow through the tide so we steamed, about an eighth of a mile apart, to the south side of the herring on a strong ebb tide. The echometer (Elac) was showing the top of the shoal laying about 10 to 12 fathoms, down to about 25 fathoms. The depth of water we would have to work in was about 40 fathoms, shallowing to 28 fathoms. The nature of the shoal and the weather conditions being taken into consideration, we thought it was policy to use the smaller midwater trawl.

I brought the *Aquarius* into position on the south side of the shoal. We were about a quarter of a mile away, giving adequate time to position the net before coming up on the herring – this distance, of course, varies according to: (1) the amount of warp you intend to shoot, (2) the strength of tide and (3) the depth of water – as you do not want to snag the net at the start of the tow. While we were steaming down-side we had discussed the warp required and had decided on 125 fathom aft, with the marks square. Turning to starboard we started to run out the net over the stern as the vessel came on to her towing heading. (In heavy weather you would shoot either before or through the wind to alleviate damage to the vessels.) The *Aquila* quickly came along our starboard side and we transferred his wing of the net by means of a heaving line. His crew secured the sweep ends to his top and bottom warps using 'G' links. Lowering the big weight into the water, we opened out the vessels so spreading the net out to see that it was clear.

Now we were ready; we let go the winch brakes simultaneously with *Aquila,* increasing our speed to 1,000 rpm, slowing down while running to the last 25 fathoms so as not to strain the warps and to avert stress to gear and boats. During this operation, we put out the decklights as soon as possible before approaching the herring as they are scared by light. Once the boats were taking the strain we started to build up the revs. from 500 to towing speed, which is about 100 revs. less than full speed. On the sonar the shoal had moved into the north west so we altered our course to get the net in a better position from north, to north west. The best part of the shoal appeared to be in that direction, but as we were entering the shoal, it moved round again into the east. We pulled round, too, into a north easterly course, with the boats about a tenth of a mile apart. Both vessels were now getting good marks, so we towed through the shoal at one eighth of a mile apart, the instruments showing the fish generally ranging from 10 fathoms down to 25 fathoms. We watched the information given on the sonar and perhaps adjusted a point to accommodate the movement of the shoal, till after about twenty minutes we came to the outer edge. We were now picking up a shallow peak on the sonar, so we towed just long enough to take our net through the fish and started to heave as soon as we entered the shallow water, to save the net. We were worried now that we might 'snag' on the hard and steep ground. We heaved as quickly as possible at about fifty turns a minute on the winch. As we did so the two boats came together again until the sweeps approached the gallows. The next step was to tow slow ahead and position the boats before the wind and sea. When the bag broke surface the two vessels then came astern simultaneously thus taking the weight off. *Aquila's* crew now unshackled their end and we put both sets of sweeps on to the power block by the heaving line method. We work one dog-rope (some work two), which we untied and put on to the winch. This done, we came astern and the crew hauled the net. To stop the fish escaping, we have a choker at the top of the herring bag, which is put into effect by the dog-rope. By this time the wind was freshening. Once the bag was alongside and strapped up on the aft starboard side, we saw we had an estimated 40-50 tonnes of herring. Next we heaved up the cod-end, carefully because of the weight, and clipped it in the Gilson. We then swung aboard and released the cod-end and about 12-15 units (1 unit = 100 kilos) spilled into the pond. We tightened the cod-end once again and let it

down into the water, ready to repeat the process until the bag was completely emptied – but unfortunately after twenty lifts in which we took aboard 298 units, the heavy motion and continued chafing against the boat's side caused the bag to burst, and we lost the rest of the catch. Nevertheless we'd had a good share, and now we hauled and started to repair the bag, while the *Aquila* lost no time in searching for another shoal.

Reports from pursers south-east of Barra Head suggested they were getting a good fishing, so we proceeded in that direction and *Aquila* located a good mark five miles in that position at 6 a.m. This time it was a big mark in deep water, ideal for the purse net. It was *Aquila's* turn now; she shot her net and raked in 265 units. As soon as she had pumped the herring aboard we set sail for the market – it was 8.30 a.m. We reached Mallaig at 2.45 p.m.; by then the wind had faired and it was not a bad day for January. We discharged a total of 563 units (56.3 tonnes) between us, fetching £6.20 per unit, and were back to sea, making for Barra Head by 11 p.m.

Peter Duthie: as told to J. Strachan

Midwater Pair Trawling from Small Vessels

Pair trawling in Europe and Scandinavia began with vessels of around 150 hp each and today is carried on with boats having up to 1,000 hp. The average power of midwater pair trawlers is probably around 400 hp. Recent experiments with modified gear have produced equipment and techniques that can enable boats of only 40 foot overall length (12 metres) with 125 hp or less, to trawl effectively in midwater.

During 1974 trials were conducted with two converted long-line boats in Newfoundland. The

Plate 24 *Kenure* D359

Plate 25 *Fragrant Cloud* D511

Plate 26 *Thomas MacDonagh* D509

Plate 27 *Camarose* SO555

vessels, *Starfish III* and *Dolphin I* were both 40 feet in length and were powered by 70 hp diesel engines. The fish sought were capelin which approach the coast of Newfoundland in dense quantities around the month of June, to spawn on the beaches. A four-seam trawl net with 79½ foot headline and 8 x 8 inch meshes around the mouth, was designed by Mr. Jack Rycroft for the project. The large net was made of extremely light nylon twine to minimise its towing resistance. Towing in depths up to 20 fathoms, the vessels were able to maintain a speed of 1.5 knots which was sufficient to catch capelin in large quantities. During one six-day period the boats landed 87 tons of the fish. The gear was later altered for herring, permitting an increased towing speed of 2 knots, but this was found to be barely fast enough for herring.

Similar trials were conducted in England with two inshore boats in the Irish Sea. One vessel, the *Golden Wake,* was 40 feet overall and had a 125 hp engine. The second boat, *Western Seas,* was 35 foot overall and had a 55 hp engine. Later she also had a 125 hp engine installed. Using a Boris four-seam trawl with 71 foot headline, the vessels were able to tow together successfully, and on one occasion caught eight tons of sprats during four tows within a 3½-hour period.

In both the Canadian and English trials, the gear was towed with a single warp from each vessel and not with twin warps as is normal for two-boat midwater trawling. This was because it was easier for the small boats to tow and haul a single wire each. Having very small deck crews, handling problems had to be simplified.

Using only single warps, there is more difficulty in steering and there is a reduction in the net mouth opening, but nevertheless it can still be successful. At one time, Russian fleets used this system when pair fishing with smaller trawlers. The single warp is attached to twin bridles, the lower of which is longer than the upper. The Canadian vessels used bridles of 10 and 13 fathoms length respectively, and the British vessels had bridles of 20 and 21 fathoms length. The Rycroft net had 28 lb. weights at the lower wing tips, and 50 lb. weights at the five fathom point on the lower bridles. For pair trawling with single wires the Boris net had 100 lb. weights attached to the lower bridles. To sink the net to depths of from 5 to 20 fathoms, the 40 foot vessels paid out from 35 to 100 fathoms of warp per side, in addition to the bridle lengths.

This method of midwater pair trawling may well be suitable for use by small craft in freshwater lakes as discussed in Chapter six.

Use of Bow Lines

The single warp towing arrangement is also used by small European vessels operating midwater pair trawls in the southern North Sea and the English Channel. French, Belgian and Dutch vessels, some with only 10 or 20 hp engines, have successfully pulled such trawls. Mostly these vessels fall into the 75 to 150 hp bracket. Their success is remarkable in that they have been able to apply the gear to the capture of demersal fish such as cod and whiting. Using bridles of from 15 to 25 fathoms, and with shorter warps than would be normal for herring trawling, they have towed the gear close to the sea bed to catch high swimming demersal fish.

A notable feature in the type of gear used by these boats is the bow line. This is a synthetic rope of from 10 to 25 mm diameter, depending on vessel size. Its length is adjusted according to the length of the warp and bridles. A simple formula to use in calculating how much bow line to use is to add the headline length, bridle length and warp length together, and divide the result by two. This gives a warp angle between the vessels of around 28 degrees. The use of the bow line makes it much easier for small boats to maintain the correct distance apart, and avoid over-straining the gear. Some larger pair trawlers also use a bow line when fishing for herring. In the case of bigger vessels, wire, or heavy nylon rope is used.

8   Ring Net Fishing

Floating surround nets, or boat-landing seines have been used for thousands of years. They did not achieve major fisheries importance, however, until the advent of powered vessels. Development then (in the past 100 years) proceeded rapidly, with two major types of surround net emerging – the purse seine used predominantly from larger vessels, and the ring net or *lampara* used mainly by smaller boats. Most ring net and *lampara* operations involve the use of two boats. One of the more successful of these fisheries was European herring ring netting, conducted chiefly in the West of Scotland.

Scotland's west coast contains many long narrow bays known as lochs. Unlike the Norwegian fjords, these lochs are mostly shallow. They may form part of larger stretches of sheltered waters such as the

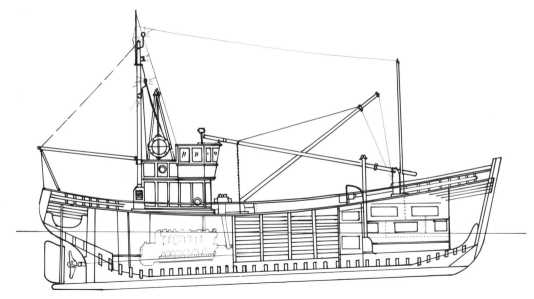

Fig 53   Typical modern ring net
Miller St Monance

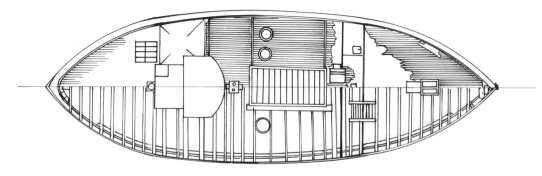

Plate 28a

Plates 28a and 28b   Types of traditional ring net vessels—two Scottish
50-foot vessels fishing off S Ireland

D M Baikie

Plate 28b

Firth of Clyde and the South Minch. Herring schools are found in there at various times in the year, and ring netting provided a quick method of catching them in quantity, as opposed to drift netting. The initial attempts at ringing herring were apparently abandoned owing to fierce local opposition, but after the First World War it rapidly gained favour. The port of Campbeltown on the Mull of Kintyre pioneered the new method, and a local man, Robert Robertson, is credited with the introduction of the first pair of boats designed specifically for the job. The vessel proved to be ideal for this kind of fishing and similar craft were soon being constructed for fishermen on both the east and west coasts. By 1938, over 30,000 tons of herring were being landed annually by ring net vessels in Britain. Ring net fishing continued to expand after World War II, as fishermen replaced their older steam or semi-diesel powered boats with modern wooden diesel engined vessels. In the north and east, seine netters were predominant, but in the south west and south east, ring net boats were chiefly in demand.

Ring net fishing was particularly suited to the supply of herring for the fresh fish market because of the speed with which the fish could be caught and landed. As early as 1923, herring which had been caught in the Firth of Clyde a few hours before, were being sold in Glasgow at 8 a.m. Drift net herring in contrast, went chiefly to curers and kipperers. A greater variety of herring was taken by the ring net which did not limit the size of the fish taken as a drift net would. At times it could also catch quantities of extremely good quality fish.

Ring Net Design

As with most surround nets, a ring net is shaped like a curtain with the ends gathered in on to side ropes and the centre part mounted slack to make the whole take up a semi-bowl shape when fishing. The bunt in which the fish are finally trapped is situated in the middle and is made of strong twine and small meshes – usually no more than 1½ inches (38 mm). The nets were originally all of cotton twine. After synthetics came on the market, the bunt part was usually made of nylon, then later nets were constructed wholly of nylon. A typical ring net could be 20 fathoms deep and 106 fathoms long, hung depth and hung length. The number of meshes was usually reckoned in scores, for instance 75 score for 1,500. The mesh sizes were counted in numbers of rows per yard. Thus a 2 inch mesh would be 35 rows to the yard, 1¾ inch would be 41 rows and 1½ inch would be 48 rows. The headline and footrope were of equal lengths and around 500 or 600 cork floats

would be strung to the head. The foot carried 200 lb. of 8 ounce (0.22 kilo) lead weights. The foot hauling line or spring rope was attached to spring lines 3 fathoms long placed 7 fathoms apart. The whole net was laid on the stern of the vessel with the corks on one side and the leads and spring line on the other. A number of buoys were attached to the headline at certain points.

Ring net boats are of a quite distinctive design and layout. They mostly measured around 50 feet (15 or

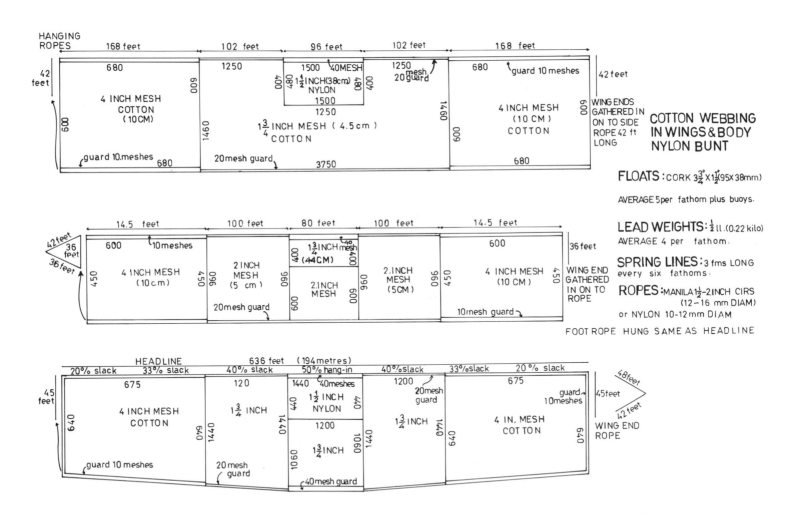

Fig 54   Ring net designs   *J Garner, F M Davies, J Grundy*

16 metres) and had a beam of 17 feet (5.2 metres) and draft of 6½ feet (2 metres). The wheelhouse is small and placed aft of midships, the foremast is usually stepped, and the side rail or bulwark is very low, often only six inches in height. The fore deck is raised a few inches and the fish room hatch, as on most herring boats, is large. The engine room is placed aft, fish hold midships, and the cabin is forward. Access to the cabin is obtained from a small hatch on the foredeck. The ring net boats were constructed with canoe (raked cruiser) or cruiser, sterns and slightly raked stems. They usually had two rubbing strakes and when rigged out, carried tyre fenders on one side, from stem to stern. Most ring net boats are varnished rather than painted and this makes them look all the more distinctive.

The multi-purpose fishing requirements on modern ring net boats call for large engine hp and powerful trawl-type winches, but originally, ring netters were extremely lightly powered. The first one

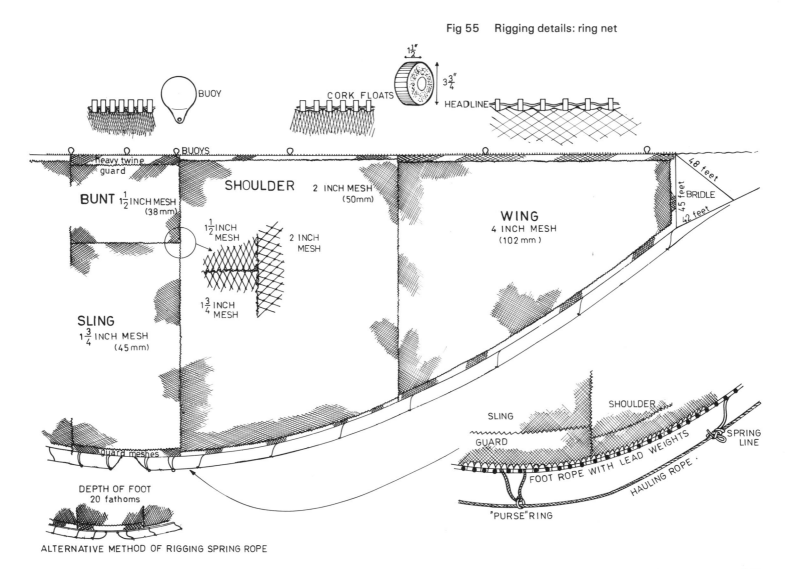

Fig 55   Rigging details: ring net

had only 25 hp. In the 1930s it was common for these boats to have 60 or 70 hp engines. In the 1950s it was still unusual for them to have any more than 100 or 125 hp. The winches used were small two-barrel gypsy type and they took up very little room on the deck. One of the main features of a ring net boat was the remarkable amount of deck space for a vessel of its size. The boats were also very manoeuvrable and quite fast for their size.

Ring Net Operation

The vessels would proceed to sea in the late afternoon and would commence searching for herring. Before echo sounders were available, they used a feeling wire. This was simply a length of piano wire attached to a lead weight. While an experienced crew man held the wire, the vessel would steam slowly through the water. Herring striking the wire would cause vibrations which the crew man could

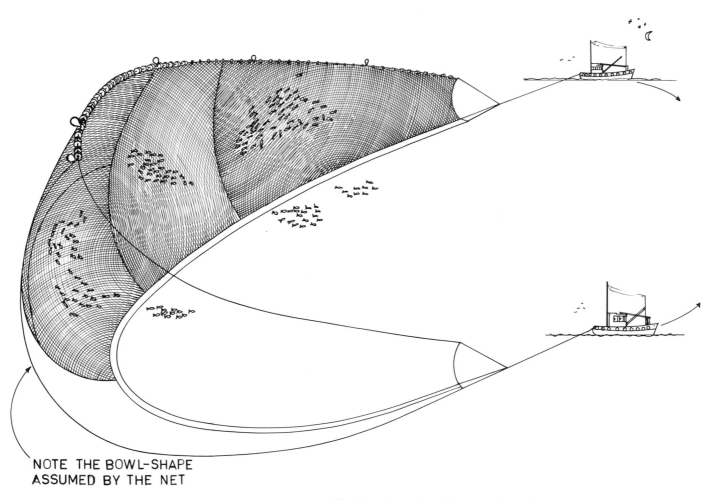

NOTE THE BOWL-SHAPE
ASSUMED BY THE NET

Fig 56   Operation of ring net: shape in water

Fig 57   Operation of Scottish ring net

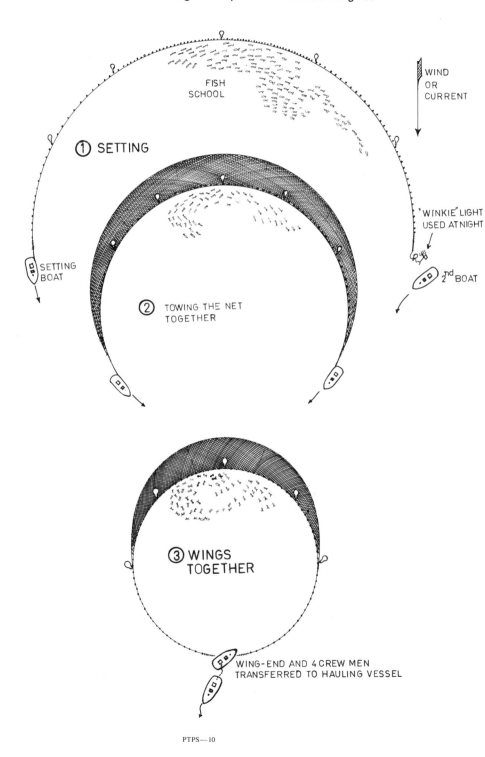

FISH
SCHOOL

WIND
OR
CURRENT

① SETTING

"WINKIE" LIGHT
USED AT NIGHT

SETTING
BOAT

2ⁿᵈ BOAT

② TOWING THE NET
TOGETHER

③ WINGS
TOGETHER

WING-END AND 4 CREW MEN
TRANSFERRED TO HAULING VESSEL

feel. He could tell the density of a school by the nature of the vibrations. Like drift net fishermen, herring ring net fishermen were renowned for their knowledge of fish behaviour and the effects of wind, tide, moon's phases and temperature, on the herring schools. Like driftermen they also worked mostly at night.

After a school was detected, one vessel would signal her intention to shoot. A marker buoy and light were thrown over with the first end of the net, and the boat began to set the gear in a semi-circle (not in a full circle as with a purse seine). The net was shot into the tide, that is, so that it might be pulled with the current. While the first boat was setting the net, the second vessel would pick up the first end of the net, or its bridle rather, which was attached to the lighted dhan buoy. Once the net was set, both vessels commenced towing slowly down-tide. This towing and closing operation is one of the ways in which the ring net operation differs from that of the purse seine. Towing could continue for some time but if the fish were properly surrounded, it would last only a few minutes, till the two vessels came together. This was a critical point in the operation. The second boat came on the outside of the hauling vessel. A crewman would quickly make a line fast while the others passed the net bridle to the hauling boat. Four of the crew men would transfer to the hauling vessel at this point, leaving only the skipper and one deckhand on the other. The deckhand would let go the mooring line and take the end of a towing spring rope which he fastened aft. Then while hauling commenced on the first vessel, the second boat would keep towing it off the net.

The two ends of the spring ropes were led to the whipping drums of the ring net winch through a roller lead on the rail. Each crew man from the two vessels had a specific job to do while hauling and two men were allocated to pull in the spring ropes and untie the stopper knots as they came up. The skipper would re-tie the stoppers on the after wing as they came off the winch. Two more men were appointed to pull in the cork line, one forward and one aft. Three crew men would draw in the webbing on the forewing and four men would perform this task on the aft wing of the net.

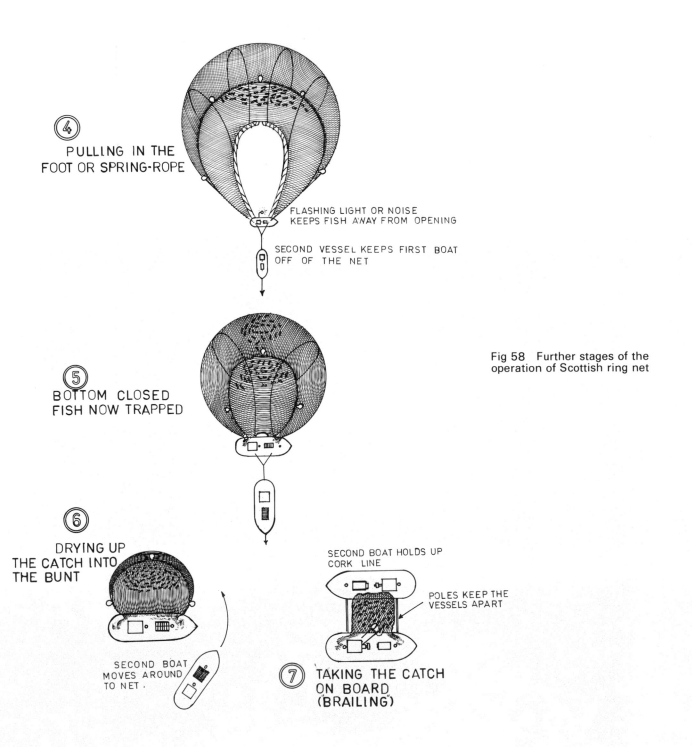

④ PULLING IN THE FOOT OR SPRING-ROPE

FLASHING LIGHT OR NOISE KEEPS FISH AWAY FROM OPENING

SECOND VESSEL KEEPS FIRST BOAT OFF OF THE NET

⑤ BOTTOM CLOSED FISH NOW TRAPPED

Fig 58 Further stages of the operation of Scottish ring net

⑥ DRYING UP THE CATCH INTO THE BUNT

SECOND BOAT MOVES AROUND TO NET.

SECOND BOAT HOLDS UP CORK LINE

POLES KEEP THE VESSELS APART

⑦ TAKING THE CATCH ON BOARD (BRAILING)

While the hauling progressed the skipper would maintain a careful watch on the gear to make sure the net was coming in properly. He would direct the crew to increase their efforts if any part of the net was lagging behind. It was important that the sole or foot of the net came in ahead of the rest. To discourage fish from swimming through the gap at the vessel, the crew would make as much noise as possible, sometimes beating on the vessel side with a large wooden mallet. Later, a flashing underwater light was used to the same effect.

Once the spring ropes were in and the sole of the net hauled up, all of the crew men would gather in the wings and shoulders of the net, leaving the fish concentrated in the bunt. The partner boat would have stopped towing by this time and would have moved round to the port bow of the hauling vessel to pick up the cork line. Some of her crew would

transfer back aboard at this stage. If the catch was heavy, fish might spill over the headline, so it was secured over the rail of the second vessel. The crew of the partner vessel would keep the boats apart using long wooden poles till the fish were all taken on board.

The herring were scooped out of the bunt by means of a brail net. This comprised a long stocking shaped bag attached to an iron hoop on a wooden pole. One crewman would dip the hoop into the bunt and others would pull it through the fish and up again, using a hauling line, so that several crans of herring were in the brailer stocking. (1 cran = 180-190 kilos approx.) The mouth of the hoop was then placed over the hold hatch or deck scuppers, and the tail of the bag was hoisted from the derrick. In this way the fish spilled through the brailer and into the hold with the minimum of pressure being brought to

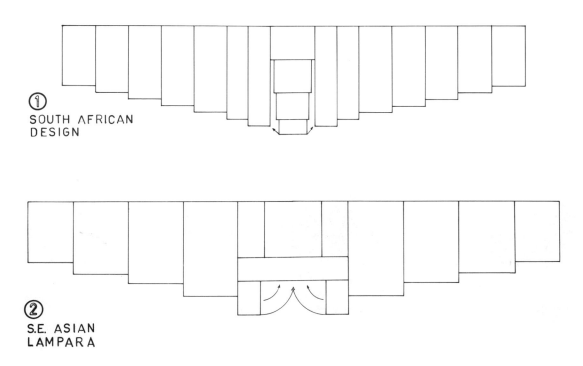

① SOUTH AFRICAN DESIGN

② S.E. ASIAN LAMPARA

Fig 59   Two-boat lampara net designs

bear on them. One or two crew men would distribute the fish around the lockers in the fish room so that the vessel maintained an even trim.

The time taken by the two experienced crews to shoot the net and haul it with a catch of about 50 crans (nearly 10 tons) might be only one hour. In difficult weather conditions, it would naturally take longer. Once all the fish were on board, the vessels separated, and the hauling boat crew relaid their net aft in preparation for another set. As both boats carried nets, they could make sets alternately.

Fishing continued through the night as long as fish were around, or until both vessels were loaded. In the hey-day of ring netting, it was not unusual for pairs of 50-foot vessels to come ashore literally 'loaded to the gunwales' with herring. These fine little craft could carry up to 200 crans (40 tons) of herring. It is a tribute to their stability and seaworthiness that hardly any ring netters were ever lost despite the large catches they carried.

Ringers rarely operated far from shore. Sometimes they fished very close to land as in a case

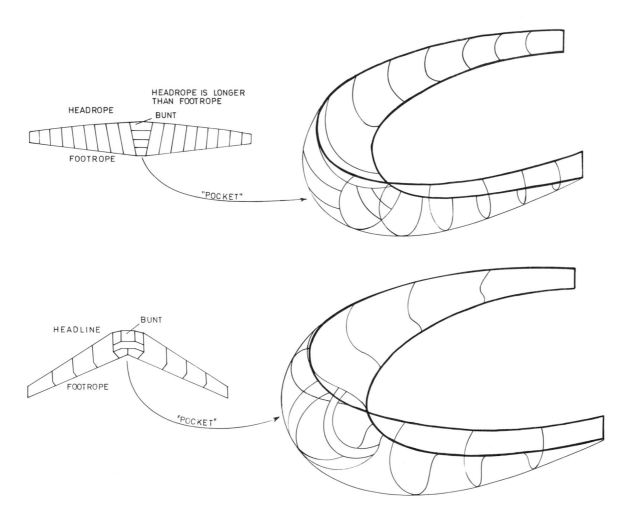

Fig 60   Lampara net forms a central pocket when rigged

112

of a set towards the beach. When fish are located near the shore, the net is set parallel to the beach, on the outside of the school. The ringers then tow the net towards the beach, only turning towards each other when they are barely a 'stone's throw' off. Considerable skill, judgement and knowledge of local waters is required to operate in such close proximity to land. Skill is also needed when fishing on shallow banks off shore where strong tidal streams may add considerably to the hauling problems.

Ring netting flourished in Scotland from 1925 to 1965 and formed the basis of many prosperous fishing communities. Most of the fishing ports in the Firth of Clyde owe their existence to the ring net, as does Mallaig in the South Minch, Britain's premier herring port. When drift net fishing was becoming a risky venture and steam drifters too expensive to operate, ring netting was still on the upsurge. Since the mid 1960s it has declined in importance with the development of midwater trawling and purse seining, but ring net boats still operate as multi-purpose vessels, which may use seine nets, prawn trawls or pair trawls at times when the ring net is less appropriate.

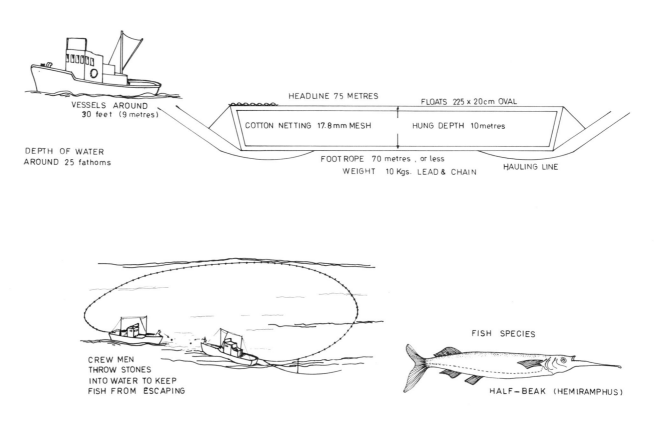

VESSELS AROUND
30 feet (9 metres)

DEPTH OF WATER
AROUND 25 fathoms

HEADLINE 75 METRES

FLOATS 225 x 20cm OVAL

COTTON NETTING 17.8mm MESH

HUNG DEPTH 10 metres

FOOT ROPE 70 metres , or less

WEIGHT 10 Kgs. LEAD & CHAIN

HAULING LINE

CREW MEN
THROW STONES
INTO WATER TO KEEP
FISH FROM ESCAPING

FISH SPECIES

HALF—BEAK (HEMIRAMPHUS)

Fig 61   Early Korean two-boat ring net for small vessels

Ring Net Fishing Development

Dual purpose fishing has characterised ring net operations since 1950. When the herring were not in season, they engaged in Scottish seining or in trawling for prawns. It was due to the demands of multi-purpose fishing that ring net boats became larger and more powerful. Some modern ringers are over 65 feet in length and have over 300 hp. They may be equipped with power blocks and are also capable of engaging in midwater pair trawling. Like most modern fishing vessels, they carry a formidable array of electronic equipment. Despite that, however, their skippers still rely greatly on personal judgement and knowledge of fish behaviour. Often they have to distinguish between schools of herring and schools of mackerel, and know how to catch the one and avoid the other.

The operation of a modern pair of ringers is not unlike that of the traditional craft. The modern vessel carries a larger net of nylon, but its design is much the same. Power blocks may assist the hauling operation. Sometimes two double-sheave pedestal type blocks are fitted, one forward and one aft, for hauling in each wing. Normally ring net boats always keep their herring in bulk, but if particularly high quality fish are being caught, they may be boxed. As ringers usually land their fish each

morning, the catch is very fresh – often only a few hours old. One feature for which ring net boats have often been renowned is their cleanliness – fishermen take a pride in keeping their craft like yachts. This is not easy to do when one is constantly working with scaly, spawny fish, but the ring netters usually succeeded in maintaining excellent standards.

Scottish type ring netting spread to England and briefly to Ireland. It was recently applied with good success to the Canadian capelin fishery. Wherever schooling fish are found, in depths up to 20 fathoms from the surface, the ring net could be used to good effect. It is applicable to small low-powered vessels, and can be used in shallow or restricted waters in which it would be difficult to use a purse seine. Ring netting can be effective in catching mackerel, pilchard, sardine and anchovy, as well as herring and capelin.

Since 1948 there have been sporadic attempts to introduce the ring net to the English Channel fishery off Cornwall, both for pilchard and for mackerel. The disappearance of the pilchard stocks from that area reduced the interest in the gear, but it is still being used occasionally for mackerel which must be landed in prime condition for the local markets. The typical Cornish crabbers and trawlers do not make ideal ring net vessels, and this has probably

Plate 29   Typical Scottish east coast ringer   *D M Baikie*

Plate 30 Ringers brailing their catch in the Firth of Clyde

restricted the spread of the gear there.

By the end of 1975 there was only one pair of ringers still using the gear regularly in Scotland. The other former ring net men had adopted the pair trawl and the purse seine instead. Many pairs of ringers could have continued to operate profitably at ring net and prawn trawl fishing, had their areas of operation been declared off limits to larger vessels. But when it became possible for large powerful pair trawlers and purse seiners to fish in the sheltered lochs and bays of Scotland's west coast, then the day of the ringer was over. It had served the fishing communities well for over forty years. But the ring net men had in the end to go in for larger vessels and more deadly gear. Among the skippers who turned to purse seining was Bert Andrew of the

Pathfinder. In 1971-72 Skipper Andrew had helped improve the ring net operation on his vessel by using deck mounted pedestal-type power blocks. The system worked well, but within two years Skipper Andrew was at the wheel of a much larger vessel, taking huge catches of herring with the purse seine. The Manson family of Mallaig also adopted the purse seine, and the ring net men of the Moray Firth and the Firth of Forth turned to midwater pair trawling instead. So the ring net ceased to be a fishing gear of importance in Scotland. Nevertheless, during fifty years of ringing herring with small vessels there was no more efficient method and few fishermen so skilled as herring ringers.

115

Lampara Ring Nets

Two-boat ring nets are used in several other fisheries. In South Africa and South East Asia, the gear is known as a *lampara* net. *Lamparas* differ from ring nets in that they have more tapered wings and a 'pouch' type bunt which is formed by a special arrangement of the central panels of netting. This may result in the net having a footrope which is shorter than the headrope. The pouch formation is like a trawl net square in reverse. It helps to prevent the downward escape of fish from the net.

There is a two-boat *lampara* operation in the Philippines which is almost identical to the ring net operation discussed, except that the towing periods may be much longer. It is used for bream-like fish which are found close to the surface in small scattered schools at certain times of year. Like many other *lampara* operations, it has declined in recent years.

Another two-boat ring net operation in East Asian waters involves the use of a surround net for half-beaks, small surface schooling fishes. This net is operated in Korean waters by vessels as small as 30 feet with only 10 hp engines. Two such vessels carry the net half on each boat and approach the school together. On reaching the fish, they separate and set the gear in a circle. When they come together, the crew begin to haul the net in by hand while some crew members throw stones into the water to drive the fish away from the opening. Once the footrope is hauled up, the fish are trapped and they are dried up into the bunt from where they are scooped into the boats. This fishery takes place in shallow inshore waters. With the advent of mechanised units with large synthetic nets, such numerous inshore seining methods are gradually disappearing from use in Asian waters.

Two-boat ring netting has largely been replaced by pair midwater trawling which is more convenient for the powerful multi-purpose inshore vessels in modern well-developed fisheries. It still has an excellent application, however, in developing fisheries with low-powered craft, or where fuel costs prohibit increased power. Wherever schooling fish are found near the surface, particularly in the shallower inshore fisheries, the two-boat ring net operation can be used to good effect.

116

Plate 31a–d   Four ringers of various size   *J Murray*

Plate 31a   A 55-foot ringer trawler OB 99

Plate 31b   60-foot ringer trawler CN 187

Plate 31c   63-foot ringer trawler SY 141

Plate 31d   Transom-sterned ringer trawler BA 25

Fishing Report: Ring Net, Ireland, January 1956

The tiny harbour at Dunmore East was crammed to capacity with local trawlers and seiners plus some Dutch herring luggers and north Irish trawlers. The two little yacht-like varnished boats putting to sea looked almost insignificant amongst the motley collection of larger vessels. It was late afternoon and the sea conditions were good despite the time of year, the clear frosty weather keeping the winds down. The two 50-footers headed out of Waterford Bay and east towards the Hook where, from that promontory to the nearby Saltee Islands, lay the most productive herring ground in all Ireland. As they sailed out, the ringers passed several trawlers and bottom seiners, their deck pounds still full of herring. Right through the daylight hours, these boats had been dragging their nets through spots of herring wherever they could be located. Offshore, beyond the three mile limit, huge steel trawlers from Continental ports were dragging their heavy bobbin-rigged gear over the bottom. To the west of that fleet, a couple of foreign drifters were preparing to set their nets in towards Helvick Head. Incredibly as it seemed, all of these vessels were getting a share of the remarkable stock of herring that came into the south Irish coast to spawn. But this season the champion fishers proved to be the little ring net boats which had come south the previous year from their native Scotland, in response to an invitation from an Irish company.

The leading vessel with the attractive name of *Arctic Moon* was skippered by Mattie Sloan, one of a remarkable family of herring fishers from the Firth of Clyde. Like her partner vessel, she was a traditional double-ender with a cruiser stern and bulwarks that rose only a few inches above the deck. Measuring 50 feet in length and with 16 feet of beam, the vessels achieved a speed of over 9 knots with their 95 hp Gardner engines. The second vessel, the *Elizmor,* was skippered by Eddie McEwan, another Clyde fisherman who was reputed to be expert at detecting fish schools with the 'wire' which was used before paper-recording echo sounders were

117

installed. In the tiny fo'csle cabin, the crew were relaxing while the cook cleared away the dishes from the evening meal. On deck, the gear was all ready for setting and to look at the spotless pounds and empty fish hold one would hardly have guessed that each boat had landed nearly 100 crans of herring that day.

The sun was now setting on the horizon to the south west and the sea surface became glassy as the last rays were reflected. A slight breeze from seaward ruffled the surface occasionally but otherwise there was only a gentle swell as the boats arrived on the grounds and began a zig-zag searching pattern about half a mile apart from each other. The echo-sounders clicked away in both the wheelhouses while the skippers scanned the surface for any tell-tale sign of a rising school of herring. A gannet dived from about sixty feet and plunged into the sea beyond the *Elizmor*. Skipper McEwan turned his vessel slightly to port and looked intently at the sounder recording. Meanwhile Mattie Sloan had circled a few marks with the *Arctic Moon*, but they proved to be scattered spots and lying rather deep. He resumed the searching course and looked for his partner vessel steaming somewhere between him and the Hook lighthouse. Darkness was now falling fast and he was about to alter course off-shore when the radio crackled and his partner's familiar voice was heard. The *Elizmor* was shooting its net.

Skipper Sloan nodded to his mate who called the crew to stand by. The lights of the partner vessel could be seen ahead and slightly to port of them the 'winkie' light bobbed up and down marking the end of the net. The vessel slowed down and the engine whined astern as they approached the net. A boathook was thrust into the water and in seconds the bridle was on board. Little was said by the skippers over the radio, but from long experience they knew precisely what action to take. Skipper Eddie did say he had set on a 'reasonable' mark. By now both vessels were towing the net, and were beginning to turn towards each other. The crew of the *Arctic Moon* prepared to board the *Elizmor*. Suddenly they were alongside and so many things were happening it was difficult to follow all the action. The bridle was passed to the *Elizmor*, the ends of the spring rope were laid to the winch, and the tow line was passed from the *Arctic Moon*.

Soon the net was being hauled in and every hand strained to pull it aboard. Once the spring rope was up, all eyes were on the bunt as the crew pulled on the netting. In no time the silver catch glistened under the water and the crew felt the weight of the fish. The cork line was pulled up on to the *Arctic Moon,* and brailing started. 'A good fifty cran' the crew reckoned, and sure enough there were 200 baskets of fish taken on board.

The vessels separated and the *Elizmor* crew had hardly time to overhaul their net when the *Arctic Moon* started to shoot. Skipper Sloan had spotted an even larger school of fish. The *Elizmor* sped across and picked up the first end of the net. Neither vessel wasted much time towing the gear and soon they were together, hauling the net which contained over 80 crans of herring. Skipper McEwan steamed off to search for more fish as soon as the last of that catch was aboard the *Arctic Moon*. Several other pairs of ringers were in the vicinity now and the skippers reverted to signalling each other by lamp rather than using the radio.

It was a little while before another set could be made owing to the activities of the other ringers, but a school was eventually located further to the east. The *Elizmor* flashed to attract her partner and the operation recommenced. Fishing continued through the winter night with hardly a rest for either crew and none at all for the skippers. By the time the sun rose, both vessels were heavily laden with fish.

A fresh breeze was blowing as the boats approached the pier at Dunmore East. Most of the trawlers and seiners had already left and there was room for them to berth with ease. *Elizmor* had taken over 150 crans while the *Arctic Moon* was carrying almost 200. Her deck amidships was almost awash. The fish were typical of the best of the season, fully mature and oozing with spawn. The crews prepared to start the work of discharging the catches, basket by basket. If they made good speed they might even snatch a half-hour's sleep after scrub-down, before they headed out to sea again.

9 Two-Boat Purse Seining

Purse seining first developed in the United States in the 19th century. From there the technique spread to the fisheries of Japan and Europe. In those sailboat days, the net was laid out from two row-boats, and all the hauling was done by hand. Until 1940, the most purse seine fishing was conducted in inshore waters using what was known as the 'two dory system'. This was the case in Iceland, Norway and Japan, in addition to the United States. After the Second World War, considerable strides were made in the mechanisation of fishing boats, and the introduction of electronic equipment and synthetic

Fig 62 Development of purse seining techniques

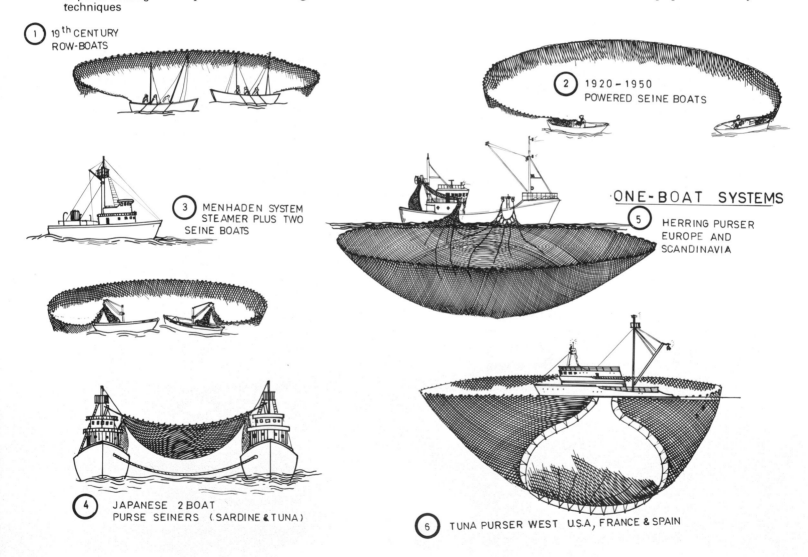

1 19th CENTURY ROW-BOATS

2 1920–1950 POWERED SEINE BOATS

3 MENHADEN SYSTEM STEAMER PLUS TWO SEINE BOATS

4 JAPANESE 2 BOAT PURSE SEINERS (SARDINE & TUNA)

ONE-BOAT SYSTEMS

5 HERRING PURSER EUROPE AND SCANDINAVIA

6 TUNA PURSER WEST U.S.A, FRANCE & SPAIN

119

nets. These factors, combined with the scarcity of fish on inshore grounds, forced the development of new purse seining techniques. So, single-boat pursing became the norm in Scandinavia, Europe, and the Western United States. Two-boat pursing continued, but it took two different forms. One could be called the small-boat/mothership operation such as was used in the American menhaden fishery and the Japanese live-bait seining for large skipjack vessels. The other system involves the use of two large boats, and it is almost exclusively confined to Japan where it is widely used in the sardine, mackerel and tuna purse seine fisheries.

We will examine both the menhaden operation and the Japanese two-boat purse seine systems in this chapter. The economics of the menhaden operation in particular, will be of interest to developing fisheries where sea distance and fish transportation pose a problem.

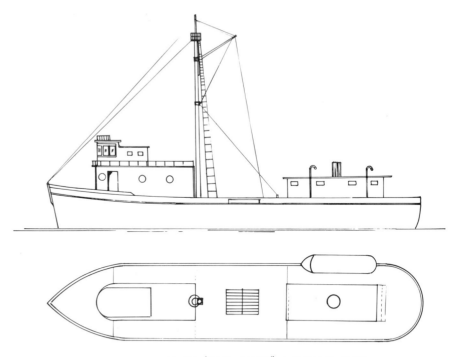

MENHADEN "STEAMER" MOTHERSHIP

The Menhaden Two-Boat Purse Seine

In the Gulf of Mexico and along the eastern shores of the United States as far as Long Island, there is found an oily, bony, herring-like fish. The Latin name for this species is *Brevoortia (tyrannus, patronus, etc)*, but most American fishermen refer to them as 'pogies'.

They are used chiefly for reduction to fish meal and oil. It is said the American Indians used to eat them, and recently Russian trawlermen were quoted as having enjoyed them as table fish, but in the United States the menhaden are caught for industrial purposes only.

The fish are found in large schools near the surface in shallow inshore waters during a season that lasts for about six months. Fishing is carried on in daytime and the vessels commence by searching for fish. This is done by visual observation from a crow's nest in the mother ship. Sometimes small spotter planes are used to locate schools and they can save the boats a lot of search time. Once a school of menhaden has been located, fishing commences.

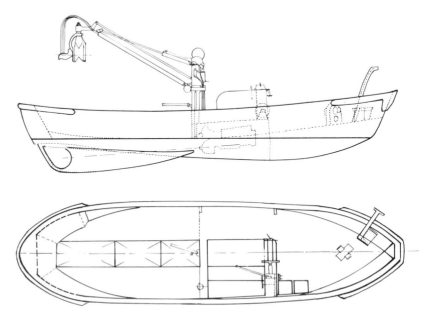

Fig 63   Menhaden purse boat with power block

The menhaden fishing is really a three-boat operation as a mothership is used. The actual fishing, however, is conducted by two small boats. Sometimes a fourth vessel, a little skiff, is also used. The two net boats measure around 36 x 9 feet (11 x 2.8 metres) and are powered by 100 hp gasoline or high speed diesel engines. They are usually of aluminium construction and carry a small pursing winch and a power block. When not in use, the seine boats are hauled up on davits and carried by the mothership. Each boat carries half of the purse seine, with the centre bunt portion slung between them. They are fast little vessels and can set the large net in around 60 seconds. They used to carry a crew of twelve on each boat, but reduced this to nine after fitting power blocks. Only three men remain on the mothership while the catcher boats are pursing.

When the mothership arrives at the location of a school of menhaden, the seine boats are lowered into the water, the crews board them, and they proceed towards the side of the school nearest the sun. Meantime another small skiff manned by one man, may stay on top of the fish to watch their movements. The seine boats remain alongside each

Fig 64   Typical Menhaden seine

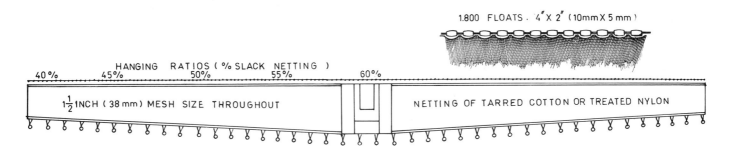

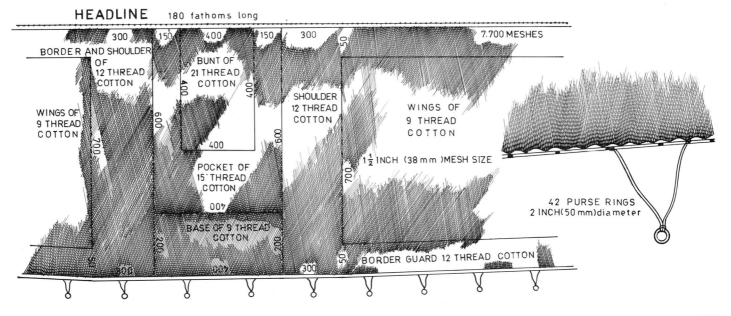

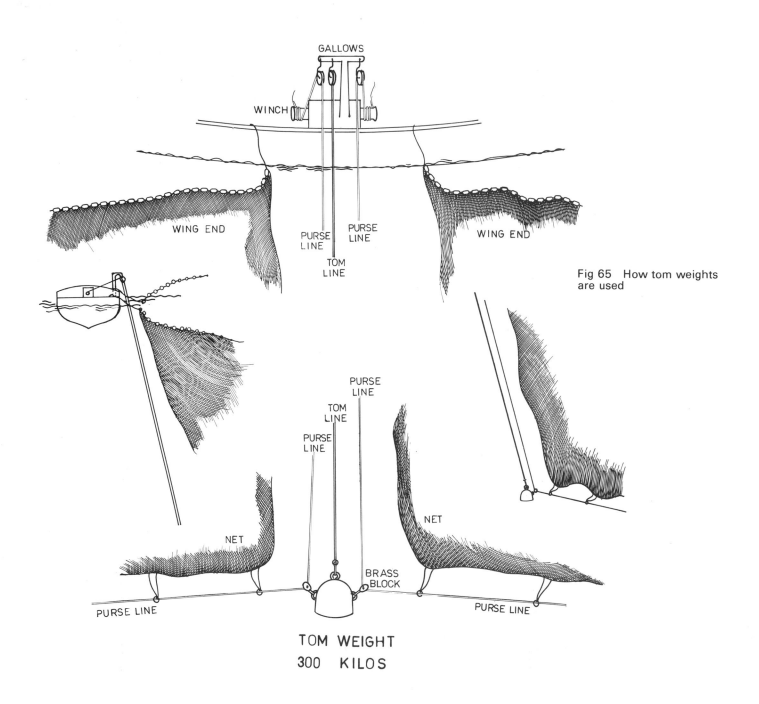

GALLOWS

WINCH

WING END

PURSE LINE PURSE LINE

TOM LINE

Fig 65   How tom weights are used

PURSE LINE

TOM LINE

PURSE LINE

NET

NET

BRASS BLOCK

PURSE LINE

PURSE LINE

TOM WEIGHT
300 KILOS

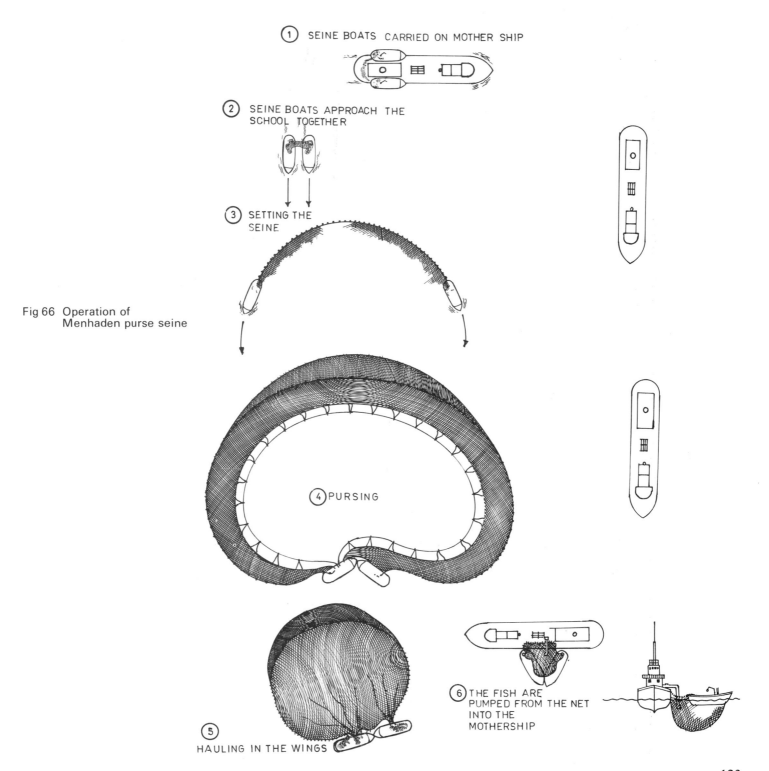

1 SEINE BOATS CARRIED ON MOTHER SHIP

2 SEINE BOATS APPROACH THE SCHOOL TOGETHER

3 SETTING THE SEINE

Fig 66 Operation of Menhaden purse seine

4 PURSING

5 HAULING IN THE WINGS

6 THE FISH ARE PUMPED FROM THE NET INTO THE MOTHERSHIP

123

Plate 32   Menhaden seiner with seine boats on board   *Bob Simpson*

other since they are 'connected' by the net that is carried half on each boat. Once they are in position, they set the net around the school, each boat turning a semi-circle. Menhaden do not swim deep, and the operation usually being in shallow water, the net may reach near to the bottom. But they have a tendency to swim towards the sun, and this is why the boats commence setting with the sun behind them.

Once the net is set and the boats come together, pursing begins. The older menhaden boats used closed rings and a purse line of Italian hemp, but modern vessels may have snap-open rings and a wire purse line. The pursing is carried on by one of the boats which is fitted with a winch for this purpose. Tom weights were commonly used on the untapered nets, to keep the net down. A 'tom' weighed around 300 kilos and had blocks fixed on each side

through which the purse lines ran. With the tom weight lowered to the foot of the net, pursing could be completed while the net walls remained vertical (see Fig. 61). Once pursing is completed, the seine boats begin to haul the net in by power block, each boat taking one wing. When the fish are concentrated in the bunt, the mothership comes alongside and the remaining part of the cork line is lashed to its rail. A fish pump is then lowered into the net and the menhaden are pumped into the hold of the ship. As soon as the net is empty, the seine boats can be taken aboard again and the mothership heads for port or searches for more schools of fish.

An average set with a menhaden seine may net 25 tons of fish, but catches in excess of 100 tons can sometimes be taken. As the carrier vessel may have a capacity of 500 or 1,000 tons it requires a fair number of successful sets for it to be fully loaded.

124

Menhaden seines were formerly made of tarred cotton webbing, but nylon nets were later introduced. They are relatively shallow; a 200 fathom net (hung length) may have a hung depth of only 15 or 20 fathoms. Mesh sizes vary with the different areas, from 1½ to 2½ inches (38-64 mm) according to the size of the fish. Usually whatever mesh size is adopted, is used throughout the net, in the bunt as well as the wings.

In recent years there was a decline in fishing for menhaden, particularly in the northern and central part of the east USA coast. This was owing to changes in the fish movements which made the seasonal fishing an even more risky venture. As the capital investment in a mothership, catcher boats, nets and a large fish meal plant, all depends on a successful menhaden season, considerable losses could be incurred if the fish failed to show up.

Plate 33   Aerial view of menhaden catcher   *John Frye*

There are several other operations which are similar to menhaden mothership-plus-two catcher boats. For some time now, Soviet and East European factory vessels have used small catcher boats which could be hauled up on davits during the long voyages to and from the fishing grounds. (Some of these 'small' catcher boats may be 50 feet long). Japanese skipjack fishing vessels require a lot of live bait for their operation, and large modern ships carry a pair of seine boats to catch the bait. These vessels look similar to the menhaden seine boats, and they operate in almost the same way. Their catch is composed of sardine-like species which the skipjack feed on with relish.

The concept of using small twin catcher boats in a fleet or mothership operation could be developed to improve the economics of long distance fisheries. Fish can be landed at a pilot station, or directly on to a transport vessel. This enables the catching vessels to make the maximum use of their capabilities, and saves them from time consuming voyages to the market. This system can be compared with the fleet operations of pareja trawlers in Spain, or herring pair trawl partnership arrangements in Scotland.

Skipjack fishing is big business in the south Pacific and Indian Ocean. When schools are located playing on the surface, the catcher vessel attracts them around her by having live bait thrown overboard. At the same time a powerful spray of water is pumped from pipes running the full length of the vessel. The combined effect of the live bait and water spray keeps the skipjack school playing around the vessel and the crew commences to catch them with pole and line gear. The hooks are barbless, and a skilled pole man can hook a fish, swing it aboard, and have his line back in the water in a few seconds. When large schools are encountered, thousands of skipjack may be caught in a few hours. It is vital to the success of the operation that the catcher boat has a good supply of live bait. This bait is caught by small boats, often using light attraction. The twin net boats must handle the bait with great care so it can be transferred live to holding tanks on the main vessel. These bait holding tanks are circulated or aerated to keep the tiny fish alive till they are required.

125

Plate 34a

Plate 34b

Plates 34a–d   Mothership and catcher vessels operating sequence showing menhaden boats and steamer   *National Fisherman*

Plate 34c

Plate 34d

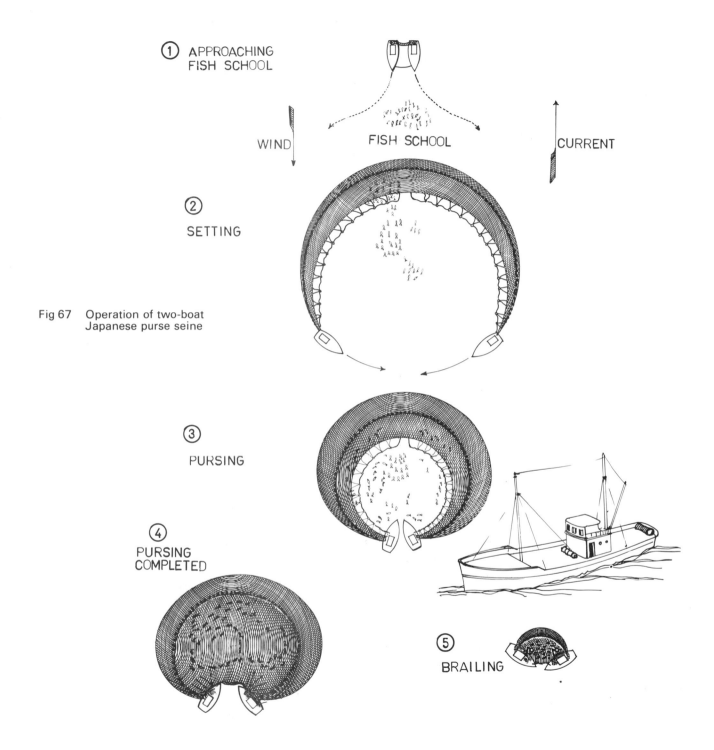

① APPROACHING
  FISH SCHOOL

WIND   FISH SCHOOL   CURRENT

② SETTING

Fig 67   Operation of two-boat
         Japanese purse seine

③ PURSING

④ PURSING
  COMPLETED

⑤ BRAILING

127

VESSELS: TWO x 80 foot 200 hp or more

HEADLINE 326 metres

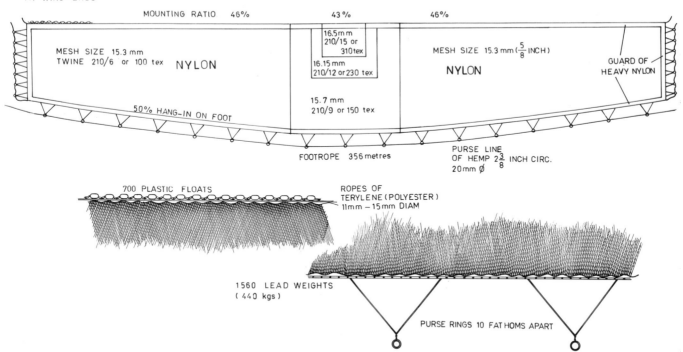

NOTE TUCKING LINE
AT WING—ENDS

MOUNTING RATIO   46%   43%   46%

MESH SIZE 15.3 mm
TWINE 210/6 or 100 tex   **NYLON**

16.5mm
210/15 or
310 tex

16.15 mm
210/12 or 230 tex

15.7 mm
210/9 or 150 tex

MESH SIZE 15.3 mm ($\frac{5}{8}$ INCH)

NYLON

GUARD OF
HEAVY NYLON

50% HANG—IN ON FOOT

FOOTROPE 356 metres

PURSE LINE
OF HEMP 2$\frac{3}{8}$ INCH CIRC.
20 mm Ø

700 PLASTIC FLOATS

ROPES OF
TERYLENE (POLYESTER)
11mm – 15mm DIAM

1560 LEAD WEIGHTS
(440 kgs)

PURSE RINGS 10 FATHOMS APART

Fig 68   Norwegian two-boat purse seine
Linde C Andersens Enke (FAO)

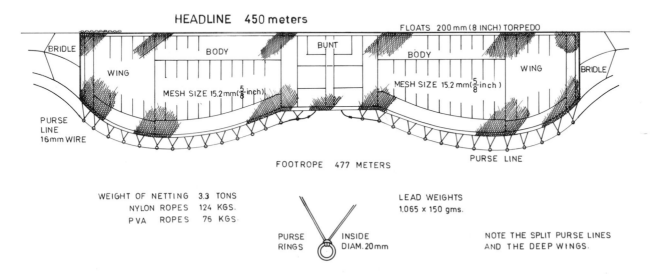

HEADLINE 450 meters

FLOATS 200 mm (8 INCH) TORPEDO

BRIDLE

BODY   BUNT   BODY

WING   WING   BRIDLE

MESH SIZE 15.2 mm($\frac{5}{8}$inch)   MESH SIZE 15.2 mm($\frac{5}{8}$ inch)

PURSE
LINE
16mm WIRE

PURSE LINE

FOOTROPE   477 METERS

WEIGHT OF NETTING   3.3 TONS
NYLON ROPES   124 KGS.
PVA   ROPES   76 KGS.

PURSE
RINGS   INSIDE
DIAM. 20mm

LEAD WEIGHTS
1,065 x 150 gms.

NOTE THE SPLIT PURSE LINES
AND THE DEEP WINGS.

Fig 69   Japanese two-boat purse seine

128

Japanese Two-Boat Purse Seining

Purse seining activities in Japan date from the late 19th century when American nets and techniques were copied by the Fisheries Agency. From the beginning in Japan, two-boat purse seining was the accepted method and it was not until after the Second World War that single-boat purse seining came into prominence. Although by 1969 there were over 200 more single-boat than two-boat purse seiners in Japan, two-boat purse seining accounted for over 46% of the seine catch, with landings in excess of half a million tons.

The bulk of the Japanese two-boat purse seine catch is composed of horse mackerel but the fishery was really built up on sardines, stocks of which have since declined considerably.

More tuna species are now being caught in purse seines and this fishery may well expand in the near future.

Sardine and Mackerel

The sardine fishery reached its peak around 1936. At this time, the vessels were still small and lightly-powered. The gear was made of cotton, hemp and sisal, and electronic aids to fish detection were not yet in use. Sardine purse seiners today work more for horse mackerel, and a few other species like anchovy and skipjack. These seiners are small, most of them today being in the 30–60 ton bracket. They carry half the net on each vessel with the mid-section hanging between. A safety line forward prevents the vessels from moving too far apart.

One distinctive feature of Japanese two-boat purse seines is the deep wings on each side of the bunt. There are two separate purse lines, the first end of each being attached to the foot at the bunt. When a school of fish is sighted, the seiners approach it from up-wind and at a given signal the safety line and net are released, and the boats pay out the net on divergent courses. The two vessels come together and attach the safety line again after completing the set, then bow to bow, each proceeds to winch in its purse line. Once the net is closed, the wings are hauled in over stern rollers rather like huge power

Plate 35a   Pursing   *Nippon*

Plate 35b   Hauling in the seine   *Nippon*

129

blocks. Some modern seiners use a power block instead of rollers.

Once the fish are dried up in the central bunt, they are taken aboard one, or both, of the vessels. A skiff may assist in the operation, but it is not always necessary.

Sardine and mackerel purse seining is sometimes used in conjunction with light attraction. In such cases, a pair of 70 or 80 foot pursers may operate with three 30 or 35 foot light boats. These light boats use both overhead and submersible lamps. The size of the generators used to power the lamps is governed by Japanese fishery laws, so no one vessel has an excessive advantage over another. Most of the light boats have generators supplying from 5 to 10 kilowatts power at 110 volts. It is possible, of course, to operate with much less. There is considerable divergence of opinion on the optimum size of light bulb for the submersible lamps. Some are 1 kW in size, some 500 W, and some less. Generally speaking, a powerful light is best to begin with to attract fish from a wide area. Later, particularly before setting the net, the wattage is reduced to bring the fish in closer to the boats. Light boats will commence attraction well apart, and after fish have been gathered, they will move slowly together until it is possible for the seine to be set around all three. Once the fish are surrounded, the lamps are

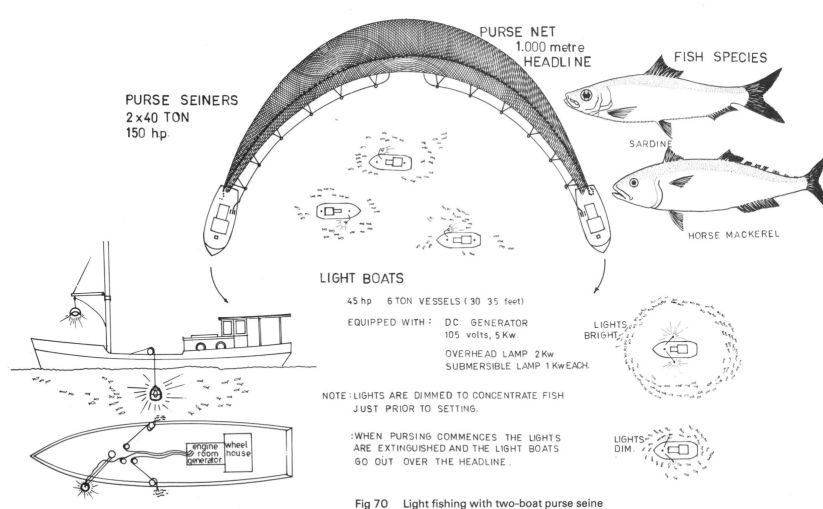

PURSE NET
1.000 metre
HEADLINE

FISH SPECIES

SARDINE

HORSE MACKEREL

PURSE SEINERS
2 x 40 TON
150 h.p.

LIGHT BOATS

45 hp   6 TON VESSELS (30 35 feet)

EQUIPPED WITH :   D.C. GENERATOR
105 volts, 5 Kw.

OVERHEAD LAMP 2 Kw
SUBMERSIBLE LAMP 1 Kw EACH.

NOTE : LIGHTS ARE DIMMED TO CONCENTRATE FISH
JUST PRIOR TO SETTING.

: WHEN PURSING COMMENCES THE LIGHTS
ARE EXTINGUISHED AND THE LIGHT BOATS
GO OUT OVER THE HEADLINE.

LIGHTS
BRIGHT.

LIGHTS
DIM.

engine room generator   wheel house

Fig 70   Light fishing with two-boat purse seine

Plate 36   Japanese tuna purse seiners   *Charles B Meyer*

Plate 37   Pair of Japanese purse seiners with the net bunt slung between them   *Nitto Seimo Co Ltd*

Plate 38   Drying the catch   *Nitto Seimo Co Ltd*

extinguished and the light-boats move out over the cork line after which they can assist the catcher vessels.

Light attraction of fish is particularly suited to purse seine and ring net fishing and it is engaged in throughout the tropics and in inland seas and lakes. Only certain fish species will respond to artificial light, and then only under certain conditions. The technique requires considerable knowledge of the local species and weather anomalies. In some cases many fish are caught because they are feeding on smaller fish attracted to the light, and not because they themselves were attracted.

Two-Boat Purse Seiners in Japan

1956-1957	Size in tons	Number	Catch in tons
Non-Powered		435	145,538
Powered			
	up to 3	41	1,474
	3-5	58	2,974
	5-10	50	8,483
	10-20	246	112,710
	20-30	102	65,666
	30-50	68	53,689
	50-100	62	24,098
	100-200	3	139
	200-500		
	over 500		
	Total	1,065	414,788

ref. Illustration of Japanese Fishing Nets

1969	Number	Catch in tons
Small type	480	238,474
N. Japan	156	292,476
S. Japan	3	7,020
	639	537,970

1969 Species Caught
(Medium and large pair seiners only)

	Tons
Horse mackerel	203,311
Sardine	75,038
Tuna and skipjack	13,434
Others	7,713

ref. T. AKAOKA   MFGW 3

Tuna

Though presently a smaller fishery in terms of numbers of vessels and tonnage of fish caught,

Japanese two-boat purse seining for tuna is better known throughout the world. This is probably because of its spectacular nature due to the large investment necessary, the global range of the fish sought, and the colossal size of the nets used. Two-boat tuna purse seines are probably the largest nets in the world. One such net may measure 1½ miles in length, and over 100 fathoms in depth. It may be composed of 10 tons of netting, 370 kilograms of mounting twine, 80 coils of synthetic rope, 4,700 floats and 3 tons of lead weights. Such a net really staggers the imagination in terms of size, to say nothing about cost! That it can be operated by two 28-metre, 550 hp boats is also surprising. Needless to say, much care and skill are necessary for successful fishing with such gear.

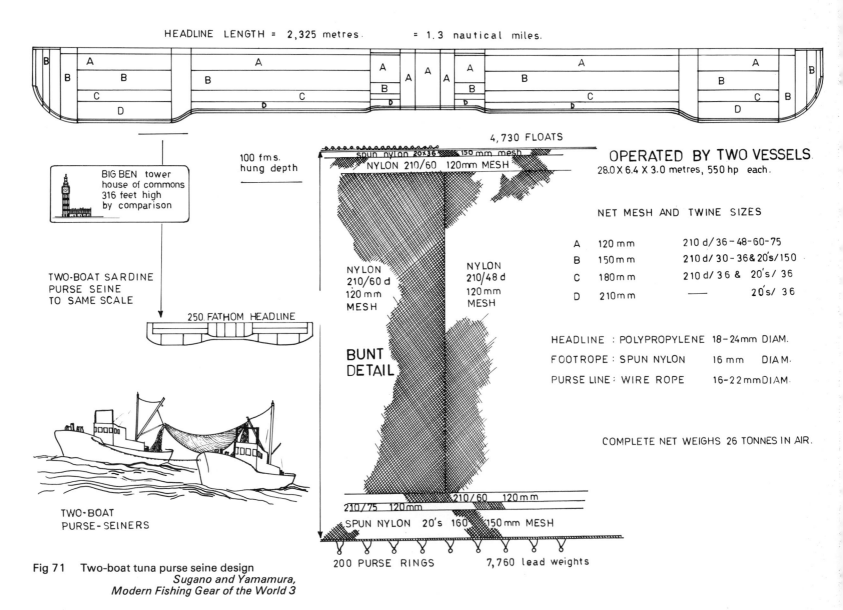

HEADLINE LENGTH = 2,325 metres. = 1.3 nautical miles.

BIG BEN tower house of commons 316 feet high by comparison

100 fms. hung depth

TWO-BOAT SARDINE PURSE SEINE TO SAME SCALE

250 FATHOM HEADLINE

TWO-BOAT PURSE-SEINERS

4,730 FLOATS
SPUN NYLON 20×36 150 mm mesh
NYLON 210/60 120mm MESH

NYLON 210/60 d 120 mm MESH

NYLON 210/48 d 120 mm MESH

BUNT DETAIL

210/60 120 mm
210/75 120mm
SPUN NYLON 20's 160 150 mm MESH

200 PURSE RINGS   7,760 lead weights

OPERATED BY TWO VESSELS.
28.0 X 6.4 X 3.0 metres, 550 hp each.

NET MESH AND TWINE SIZES

A	120 mm	210 d/ 36 – 48 – 60 – 75
B	150 mm	210 d/ 30 – 36 & 20's/150
C	180 mm	210 d/ 36 & 20's/ 36
D	210 mm	—— 20's/ 36

HEADLINE : POLYPROPYLENE 18–24mm DIAM.
FOOTROPE : SPUN NYLON 16 mm DIAM.
PURSE LINE : WIRE ROPE 16–22mm DIAM.

COMPLETE NET WEIGHS 26 TONNES IN AIR.

Fig 71   Two-boat tuna purse seine design
Sugano and Yamamura,
Modern Fishing Gear of the World 3

132

Most tuna are caught by long lines, floating long lines which may hang 100 fathoms deep or more! Small tuna species like skipjack can be caught by pole and line near the surface. The reason that tuna purse seining is relatively recent in development (the Americans pioneered it off California) is that these fish are extremely fast swimmers and though one may locate schools fairly easily, it is quite another problem to net them! Tuna seiners must be fast and in this respect the two-boat operation has an advantage in being able to set the net quickly. Two searcher-attendant boats are also used in pair tuna seining, and it is indicative of the type of fishery that these are as big and as powerful as the catcher boats themselves. They locate the tuna fish, help keep the school inside the ring during setting and pursing, and help tow the seine boats off of the gear during hauling and brailing.

Specifications of Japanese Two-Boat Tuna Purse Seine Vessels (W. Africa 1969)

	Seine Boat 1	Seine Boat 2	Search and assist Vessels	
Dimensions in metres	28.0 x 6.4 x 3.0	28.5 x 6.4 x 3.0	27.0 x 5.0 x 3.5	27.0 x 5.5 x 3.5
Gross tonnage	145.5	145.2	99.1	99.2
Main engine hp	550	550	620	620
Auxiliary engine hp	300	300	100	100
Fuel capacity tons	33	33	46	46
Fish hold capacity m³	78	78	42	44
Fish detection equipment	Sonar	Sonar	Echo sounder	Echo sounder
Purse winch pull speed	9 T a 24 m/min. 4.5 T a 36 m/min.	9 T a 24 m/min. 4.5 T a 36 m/min.		
Side winch pull speed	4.2 T a 48 m/min.	4.2 T a 48 m/min.		
Net hauler pull speed	3.5 T a 20 m/min.	3.5 T a 20 m/min.		

The sinking speed of a purse seine is important, particularly for tuna species which can dive quickly to escape capture. This is one reason why the net is set more in a pear shape than circular. The boats may sail relatively close together while shooting the bunt, and stream apart when the wings start running out. Most purse seine skippers prefer an oval shape of set, with the vessel(s) at the downwind point. This makes for less chance of fouling during pursing and hauling.

There are many species of tuna fish, most of which swim rather deep in the ocean. Skipjack, one of the smaller species, are found mostly on the surface. Yellowfin tuna can also be seen at the surface, particularly near tide rips or cross-currents. Bluefin tuna are mostly a deep swimming fish but at certain times they also can be found near the surface. It is primarily for these three species that the tuna seiners operate.

The spotter vessels stay down-tide of a school till the catcher boats begin setting the net. Once the net is set, the spotter vessels steam back and forth by the net gate or mouth, to frighten the fish away from that escape route. Once pursing is completed, they begin to pull the seine boats off the net, using long, heavy tow-lines. When the fish are dried up in the bunt section, the seine boats are pulled apart so the bunt forms a rectangular pouch between them. At this point, if the catch is heavy, a blanket net is passed under the fish from one boat to the other. This is to avoid bursting the main net with the weight of fish. The tuna are then brailed out with a triangular brail net which can remove 20 to 30 tons per hour. To save the tuna from being suffocated by the blood from the other fish crowded in the bunt, the attendant boats tow the seine boats and net slowly into the tide during brailing. This allows a flow of fresh sea water through the net.

Operating off West Africa since 1964, Japanese fishing companies have been able to land annually from 2,000 to 4,000 tons of tuna (mostly yellowfin and skipjack) per pair of purse boats in operation. The large nets are prone to distortion due to shrinkage and stretching, and they are vulnerable to damage, breaking under strain at times. A freezer-mothership attends the fleet of pair seiners, and takes their catches from them for processing and transport to market many thousands of miles away.

Plate 39   Skiff being launched to set net

Plate 40   Close up of bow line between vessels

Plates 41a–d   Composite showing the stages of closing the net ready for hauling—note the bow line

Plate 41a

Plate 41b

Plate 41c

Plate 41d

134

Plates 42a–42b   Composite showing drying up a catch of large tuna   *Nitto Seimo Co Ltd*

Plate 42a

Plate 42b

Plate 43   Seine boats in action closing the net round a school of menhaden   *Charles B Meyer*

10 The Technique of Fleet Operations

It will be obvious to readers that the success of two-boat fishing is enhanced in many cases by the group or fleet operations that many pair fishers engage in. The team-work and co-operation necessary in two-boat fishing can be applied to fleet operations to the benefit of all concerned. Thus group fishing can be looked upon as a logical extension of pair fishing operations. In many of the fisheries referred to in this book, the basic unit is not actually a pair of boats, but a fleet of vessels which may number three, five, seven or more. There are both technical and economic reasons for this. Often in technical books dealing with fishing gear and methods, such peculiarities of operation are ignored, but as they form part and parcel of the fishery, it is better to pay some attention to them. Readers with no background knowledge can be easily misled by purely technical data. After all, commercial fishing, like any other business, is a struggle for a livelihood against operating costs, inefficiency, and the natural elements. There is no sure and easy road to success, even with the best equipment in the world. Economic and social factors must be studied as much as technical aspects, in assessing the potential of any new fishery.

Since most pair fishing vessels are privately owned with the skipper and possibly some crew members holding shares in any one vessel, the particular kind of fleet operations engaged in varies from group to group. It is the result of the consensus of agreement reached by each party. For this reason one will find much variation and flexibility in the operations of different fleets. But in general they all seek the same basic economic advantage of operating as a group rather than as individuals or pairs. This loose but effective business arrangement (with no written contract!) is typical of the privately owned section of the fishing industry with its associations, co-operatives, mutual insurances, box pools and other joint activities. Company owned or State owned fleets also engage in group operations, of course, but to date the Japanese and Russians have led the way in this field. Japanese tuna fleets and Russian trawler fleets have developed well organised systems for search, capture and transport activities by their global fishing vessels. For the purposes of this paper we will consider only those kinds of fleet operations affecting pair fishing vessels mentioned in the text.

Catch Transport Operations

If one studies the activities of individual fishing vessels from any one port, one will find that the time spent actually catching fish is quite short, and that a surprising amount of time is taken up in getting to and from the fishing grounds and landing the catch and replenishing the stores. Studies of many different fisheries have shown that few small fishing vessels spend more than 200 days a year at sea. It would be unusual for any but long-distance freezer trawlers to spend more than 250 days a year at sea. Now, of the 200 days or less that a fishing boat is at sea, a large proportion of time is spent in travelling between the fishing port and the fishing grounds. Inshore fishing grounds may lie only a few hours distance from the market, but as most inshore boats land their fish daily or bi-weekly, this adds up to a lot of steaming time. In addition to the steaming time, discharging and re-storing time in port must also be considered. If there is no market or landing place within eight or twelve hours of the grounds, then fresh fish trawlers could well benefit from a fleet transport arrangement. (The benefits of sharing the transport to market apply primarily to bottom trawlers which do not normally load up in one or two days. Bulk fishing vessels like purse seiners, ring netters and midwater trawlers, may fill their holds in one night and consequently could not share transport with each other.) The Spanish pareja trawlers operated a scheme like this as have other kinds of bottom fishers.

Long distance fishers like the pareja trawlers, operating in groups of five, six or seven vessels, would put their combined catches on one boat every three or four days. Assuming the fish hold capacity was 50 tons plus ice, and the catch per boat per day was two tons, this would about fill one vessel (6 x 2 x 4 = 48 tons). The benefits of this system are obvious. If the vessels worked only as pairs, then every ten to fourteen days they would have to return to port, not just because the ice was finished or the hold full, but in order to land the fish in reasonable condition. By using the fleet method, each vessel would get around twenty days uninterrupted fishing time, and what is more, the fish would be landed regularly in really fresh condition, being only

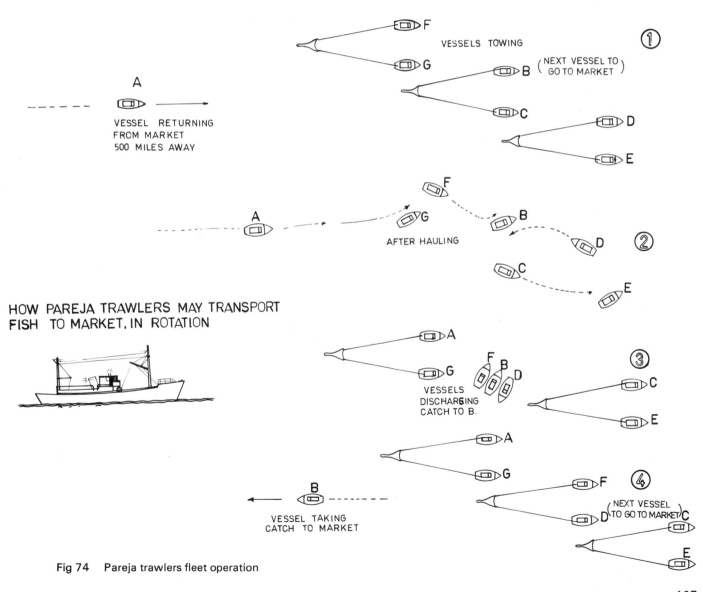

HOW PAREJA TRAWLERS MAY TRANSPORT FISH TO MARKET, IN ROTATION

Fig 74   Pareja trawlers fleet operation

4–5 days old. Passing the catch from one vessel to the other requires calm sea conditions. If the sea was rough, Spanish boats used to pass the fish in the shelter of one of the bays on the west Irish coast. The actual discharging of fish on to the vessel bound for the market did not interfere too much with the fishing operations. Of each pair of trawlers, only one would take fish on board. Assuming a fleet of seven vessels, one of which was always en route or returning from the market, that would leave six vessels or three pairs actually fishing. One vessel from each pair would have fish on board. The vessel due to go home would separate from her partner ship which would be joined by the vessel returning from port. One vessel from each remaining pair would come alongside (one on each side) and begin discharging. Their partners could then form a pair and continue fishing. Once the fish was discharged, the transporting vessel headed for market and the two discharging boats could pair up and recommence trawling. This illustrates how little loss of fishing time could be incurred. During the days when only hake fish were sought by parejas, the discharging could be done at night as they did not tow during the hours of darkness. But later when cod, coalfish and prawns were also being fished, parejas would tow by day and night.

The fleet operations of Japanese bull trawlers resemble those of pareja boats in former years, except that larger numbers of vessels are involved, and the fish carrier vessels do not engage in fishing. A fleet of 80 bull trawlers (40 pairs) operating within a ten mile radius on grounds 36 hours steaming distance from port, might be served by four or five carrier vessels. The carrier boats are usually old trawlers which have been converted to engage exclusively in transport work. Each carrier vessel spends about 24 hours on the grounds, collecting fish and shrimp from the trawlers. The catches are usually transferred while the boats are towing. The fish are packed in 25 or 30 kilo wooden boxes, with ice. Vessels making six or seven tows a day, with an average of 30 to 50 boxes a tow might accumulate 200 or 300 boxes a day (about six or eight tons). If five transport vessels were in operation, each with a capacity of around 300 tons (10,000 boxes), then

each one could take the accumulated five or six day catch from eight pairs of trawlers. The bull trawlers return to port every 35 or 40 days. If there were no transport vessels serving the fleet, they would have to land their fish every 10-14 days. The transport ships are probably saving the catcher vessels over 100 days' fishing time a year. If each carrier vessel serves eight pairs of trawlers, then it allows 800 extra fishing days per year over a theoretical two-boat operation. Thus the transport boats prove to be an extremely worthwhile investment. It is important to remember, however, in analysing these situations, that the actual operations are much more complex, with many more conflicting factors than can be clearly shown on paper. Catch rates, weather conditions, distances and other factors vary considerably over any one year, and from year to year. Nevertheless, it is still apparent that the fleet transport system can be a most profitable one.

Similar operations have been engaged in by bottom trawlers and bottom seiners operating six to twelve hours distance from a market or landing place serving fresh fish outlets. In such cases, quality is of prime importance and the fish should reach the consumers within two days of capture. A fleet of four boats would put their catches in boxes on one of the vessels each evening for four days. The transporting vessel would take the fish ashore, land it, and return to the grounds by the following morning. Each boat would make one such trip in turn. On the fifth night all four vessels would return to port for the week-end. This enabled the vessels to fish profitably and serve a high quality market, in an area where it was not feasible to operate as a single unit.

Catch transport operations of this nature are not much engaged in now by fishermen in well-developed countries. The establishment of new harbours and markets close to the fishing grounds has made it possible for inshore vessels to bring their own catches back fresh to market without detracting too much from fishing time.

In large developing countries, however, where there may not be suitable harbours or markets along many hundreds of miles of coastline, then the system can be used to good effect. It can also be applied to

Plate 44a   Plates 44a–44b   Japanese bull trawler vessels   *Kanashi Shipbuilding Co*

Plate 44b

small scale fisheries with similar benefits. The writer had responsibility some years ago, for small gill net boats operating in large lakes in Africa. Access roads were limited and the boats had to fish many miles from the fish market, so that it was not possible to set and haul the gill nets every night. By using ice, and by having each boat in turn bring the total catch back to port each day, it was possible to double the fishing time on the grounds, and also to land the fish daily in excellent condition.

One further method of catch transport deserves some mention as it is widely used by herring pair trawlers in Europe. In this case, instead of transporting the fish to the market, the carrier vessels bring the market to the fishing fleet. The transport vessels are herring 'luggers' – small cargo boats which may be converted trawlers or drifters. They measure from 30 to 70 metres overall, and can carry over 1,000 barrels of herring. Such luggers come from Norway, Germany, Holland and France, and head for remote herring fisheries such as the Shetland Isles, west Scotland, south and west Ireland, when the fisheries are in season. The luggers anchor in a bay or near the harbour used by local pair trawlers. Their representatives attend the local fish auctions where they bid on catches like other buyers. If there is no auction in the vicinity, then they might come to a private deal with the pair fishermen. The herring are then discharged directly into the lugger, in which they are salted and packed in barrels, ready for sale, or for further processing in the country of destination. The fleets of luggers which follow the herring trawlers around the coast, as they in turn follow the herring schools, perform a valuable service in expanding the capacity of local markets, and in preventing gluts caused by excessively heavy landings in a relatively short period. It should be borne in mind that the lugger system is really a post-harvest operation which is arranged by merchants and not by fishermen as such. They are only feasible in cases where the fishery is remotely situated or difficult to reach overland, and the ultimate destination for the fish is many hundreds of miles away. The practice of 'klondyking' for herring as it is called, has been pursued since the days of the sailing drifters and steam drifters. Ice was used in addition to salt to help maintain the quality of the fish.

Search and Capture Operations

The fleet operations of midwater pair trawlers differ from those of pareja trawlers in that they are primarily oriented towards the search for and capture of fish, rather than its transport to market. Midwater pair trawlers do not normally operate much farther than eight hours from the nearest market and often they may be fishing within 50 miles of the port. They are essentially 'bulk' fishing vessels as compared to bottom trawlers which land proportionately less in terms of weight. This does not mean that quality is neglected on pair trawlers — far from it — the herring are boxed, iced and stowed with care on modern vessels. Some boats are equipped with sea water tanks which are cooled by ice or refrigeration. Herring kept in these tanks or containers can be landed in perfect condition. Speed is essential in getting herring to market and it would not be feasible for a fleet of herring trawlers to put their catches on a transport vessel. It would take too long, and the vessel probably could not hold all the fish. So, apart from sometimes allowing only one vessel of a pair to load up with fish first, there is no effort by midwater trawlers to pool their resources in the transport of fish. The main purpose of their association is to achieve greater efficiency in finding and catching the fish.

Herring fishing differs a lot from demersal trawling in that there is a much greater fluctuation in the location and behaviour of the fish over any 24 hour period. The only times when herring are static in daylight and darkness is when they are spawning. Otherwise they are very active, usually forming schools which approach the surface just after sunset and just before sunrise. During daylight hours they scatter, or seek protection around rocks or in deeper waters. This results in two peak fishing periods for midwater trawlers, one after dusk and one before dawn. If these last approximately three hours each, then this gives only six hours in the day when fish may be taken in quantity. Weather conditions and the skill of the fishermen will alter the picture slightly. During moonlit nights, fish may be taken throughout the night. An experienced skipper may sometimes be able to catch herring in daylight, but this is the exception to the rule. Generally speaking, the fishermen must net their catches during the brief 'peak' period or they will lose out altogether. It is in this situation that the fleet operation scores over a simple two-boat operation.

Midwater trawlers make short tows – rarely longer than 30 minutes when the fish are schooling, and often much less. Shooting and hauling take only 15 minutes each, so the total time necessary for one tow is about one hour or less. Taking the catch aboard takes up much more time. It would take one or two hours to take on board 100 crans of herring (nearly 20 tons). The fish are hoisted on deck in lifts of ten crans (40 baskets) maximum. Excessively large lifts would crush the herring and some fishermen prefer to take only five crans or less at a time. Weather conditions and the skill of the crew will affect the speed of the operation. But however fast the crew are, the fact remains that one of the pair

trawlers is unable to continue fishing for an hour or more after a tow. If a fleet operation is being conducted then the free vessel can pair up with another boat in the group and perhaps net another bag of fish during the vital 'peak' period.

The fleet operations actually begin much earlier during the searching period. As the task of locating herring schools is vital to the success of the operation, a fleet of say five trawlers has a big advantage over a pair of boats. News of herring schools will eventually reach all the pair trawlers in the area, by radio-telephone, but as the fish do not wait around to be caught, the boats on the spot first will have by far the best chance. In order to limit information initially to vessels of their own group, skippers will talk only on special vhf frequencies, sometimes using a 'scrambler'. Occasionally code words or phrases are used. The fishing fleet will be informed, of course, but not until the group vessels have made successful tows.

A fleet of say five vessels searching an area on a

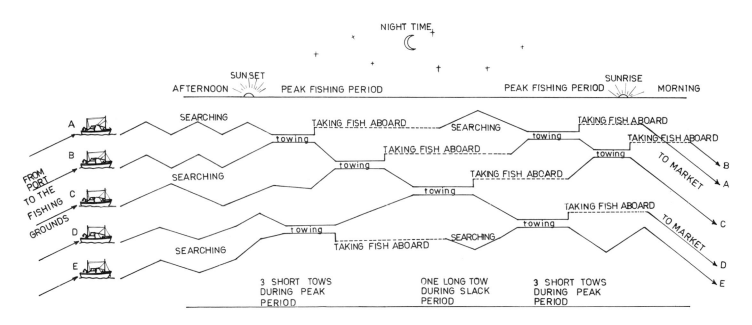

Fig 75   Midwater pair trawlers fleet operation

pre-arranged series of tracks has a much better chance of spotting fish than a fleet of only two boats. As the search period begins before sunset, the skippers are looking for the first tell-tale signs of the formation of a school. Usually the fish will begin to group fairly deep and the school will gradually increase in size and approach the surface as the sun sets.

The group method of fish hunting is also helpful during towing. One pair of vessels will set their gear and commence towing in the direction of a herring school. But the fish may have moved from the position where they were first spotted. A free vessel from the group can slowly scan the area ahead of the towing pair and thus guide them directly on to the fish. Care must be exercised not to frighten the fish by excessive engine noise and the searching vessel will avoid criss-crossing the school at full speed. Detection is accomplished chiefly by echo-sounders, some of which may be fairly sensitive high-frequency or multi-stylus models. Some pair trawlers are also equipped with sonar which can be used to good effect in following the movements of a fish school.

The economic benefits of group fishing are quite clear. It can be most profitable providing there is good teamwork and co-operation. The financial sharing arrangements vary from group to group but generally speaking, each vessel receives an equal share of the proceeds from the combined catches. The operational expenses of each vessel are then deducted from its own share, though some groups pay for all the fuel oil from the gross earnings. It is better that all the crew men in a group receive the same share in pay, but as some boats have higher food bills, travelling expenses or crew insurances to pay, it does not always work out that way. A more difficult sharing problem is the cost of fishing gear. In theory each boat should purchase the same value of fishing gear in a year but this is difficult to gauge sometimes. Skippers buy different kinds of nets, ropes, fish boxes, and other gear. To compound the problem, groups may break up during the off-season and engage in single vessel bottom trawling or seining for a period. However, despite these problems, group fishing by midwater pair trawlers is

successfully prosecuted and will probably continue to be the most economical method of tackling this kind of fishing.

Group Fishing by Scottish Pair Trawlers

One of the best known teams in the Scottish fleet was 'the big five'. It was composed of five herring trawlers from Peterhead who through skill, perseverance and co-operation have perfected the team-work and rapport essential to group fishing activities. Each skipper was a successful fisherman in his own right, but together they comprised the most formidable midwater trawl fishing team in the country. They began to operate as a team when they had wooden vessels of around 80 feet in length with 240-400 hp. By 1975 four of the older vessels had been replaced with larger, more powerful steel trawlers, and a fifth new vessel was also due for delivery. The team was composed of:

Skipper Jim Pirie, MV *Shemara* PD 78, 637 hp, 86' x 22' 6" x 12'.
Skipper John Buchan, MV *Sparkling Star* PD 137, 750 hp, 87' x 24' x 12' 3".
Skipper William Strachan, MV *Juneve III* PD 215, 850 hp, 86' x 22' 6" x 12' 6".
Skipper John Alec Buchan, MV *Fairweather V* PD 157, 637 hp, 85' x 22' x 12'.
Skipper Arthur Buchan, MV *Ugievale II* PD 105, 450 hp, 80' x 21' x 12'.

The vessels were built in different yards, but they followed the same pattern, having traditional lines and modern layout. Although they spent most of the year pair trawling for herring, they were multi-purpose vessels and could convert quickly to bottom trawling, seine netting or purse seining. Their fish rooms were laid out to carry fish in boxes, in bulk or in containers. Mostly the herring were boxed and only extra large catches carried in bulk. Boxed herring fetch better prices on the market than bulk fish. For quality, it is not possible to better the herring in chilled sea-water containers, but though they fetch top prices, the method is not used regularly as marketing arrangements usually have to

Plate 45a *Shemara* PD78

Plates 45a–45b Scottish pair trawlers

Plate 45b *Ugievale* PD105

be made in advance. It does, however, hold much promise for the future.

The big five followed the herring schools around the British Isles, fishing in the Irish Sea, around the Hebrides, at Shetland, in the North Sea and on an occasion, on the traditional grounds off east England. All five vessels participated in the search, capture and transport operations, and the gross earnings were divided equally amongst them.

They usually shared the fuel costs jointly, but each vessel would pay for the aggregate cost of its own food, insurance, crew transport and other such expenses.

Usually the fish are sold in cran units. Previously, the cran measure, a unit of great antiquity in the herring industry, was measured by filling four quarter-cran baskets, each holding approximately 100 lb. of fish (although by definition the old cran measure was one of volume rather than weight). In 1974 the Herring Industry Board introduced a new unit cran measure of 100 kilos. Thus ten of the new unit crans would weigh one metric tonne, as compared with around five and a half old crans. When boxing, the fish are packed in polyethylene HIBEX boxes which hold 25 kilos each. Thus four HIBEX boxes of herring make up one unit cran. The HIBEX system was introduced by the Herring Industry Board to provide a national, uniform and hygienic container system. The boxes measure 24″ x 14.5″ x 8.5″ (603 x 368 x 214 mm).

Pair trawler teams are not necessarily composed of identical vessels. Like the Grimsby cod trawlers, a pair may be quite unevenly matched in size, type and power. There is one trio of pair fishers that are almost identical. Three steel vessels of Tynedraft design were selected by their Peterhead skippers when renewing their fleet. They are the *Unity* PD 209, the *Starlight* PD 150 and the *Constant Friend* PD 83. Each trawler measures 86′ x 22′ 6″ x 12′ 6″ and is powered by a 660 hp Alpha engine. The *Accord* PD 90 is a steel stern trawler which pair fishes with the wooden cruiser-sterned *Star Crest* PD 114, and another traditional boat *Faithful II* PD 67. Some teams may have a purse seiner as a partner like the *Aquila* FR 105, which teamed up with the *Aquarius* FR 55, *Kallista*

FR 107 and the *Uberous* FR 50. The *Aquila* and the *Kallista,* skippered by Jim Slater and George Watt, have 850 hp engines installed, and were probably the most powerful pair in the Scottish herring fleet. Purse seiners sometimes engage in pair trawling when herring schools are scattered, and one such team is the *Vigilant/Lunar Bow* partnership. The techniques used by each team may differ slightly according to the number of vessels and the particular circumstances. But by pooling their capabilities and co-operating in the search for, and the capture and transport of the fish, they are proving the value of fine teamwork and organised fleet operations in the herring fisheries.

Long-Distance Operations with a Transport/Mothership

Some fleet operations by two-boat purse seiners off Africa and Asia involve the use of both searcher vessels and transport ships which may also act as motherships to the fleet. This kind of operation is made necessary by the great distances between the fishing grounds and the market, and the lack of any local harbours sufficiently well equipped to act as bases for operations. The absence of harbours or fuel depots, ice factories or engineer shops, makes it difficult to prosecute fisheries in remote areas. This is particularly the case in vast stretches of the ocean off remote parts of tropical and sub-tropical continents, or around isolated oceanic islands.

The problems of global distance fishing have been tackled by the state-owned fleets of the USSR and east European countries, and by large Japanese fishing companies. Generally speaking Russian fleets have concentrated on trawling but a few of their motherships carry small purse seine vessels. Japanese fleet operations involve many types of fishing including long lining for tuna, drift netting for salmon and purse seining for sardine and skipjack.

The Japanese two-boat purse seine fishery for tuna is an example of fleet operations involving the use of catcher boats, searcher vessels and a mothership which also processes and transports the catch. The high degree of specialisation necessary in such an enterprise makes it financially impossible for any but the largest of fishing companies or state enterprises to engage in. The mothership must be equipped, not just to handle and process the catch, but also to carry fuel, ice, stores and spare parts for the smaller ships, and to carry out repairs in the absence of an adequate marine engine or electrical workshop ashore. The mothership remains in close radio contact with each of the catching and searching units. To obtain maximum use of the vessels during the fishing season, it is imperative that their operations continue with the minimum of stoppages through bad weather, breakdowns or other adverse occurrences. This can be very hard on the crew members who are away from home for several months at a time, and the mothership is usually equipped to provide them with some form of relaxation and entertainment.

During periods of heavy fishing, the mothership may have difficulty in coping with large quantities of fish in a short period of time owing to the limited capacity of the processing facilities. At other times the shipboard 'factory' may lie idle for days. Because of this problem it would appear better to have several small fishing units rather than a few large ones serving the factory vessel, but the nature of the fishery must also be taken into consideration.

Mothership operations in themselves are not new. For centuries now, Portuguese fishermen have sailed to Greenland in their large schooners to fish for cod from little two-man dories. Dozens of such dories, powered only by oar and sail, spread out over the ice-cold waters around their stately mothership, and commenced fishing with their little hand-lines. They returned each evening, loaded with cod-fish, through the treacherous fog-bound seas, to the schooner where the fish were loaded, dressed, washed and salted. One fishing trip lasted for five or six months from about May to September. Although it must be one of the most rigorous fisheries in the world, it was probably the most long-established mothership type operation.

Another type of search, capture and transport operation using two-boat purse seines, is the light-fishing for sardine and mackerel conducted in Asian waters. This is a much less complex system than the tuna seining operation, and it also requires much less capital investment. A complete fleet unit could be formed by private fishermen under a co-operative organisation. In this case the motherships transport the catch, but do not process it. They may even use only ice to keep the fish fresh, if the fish market is less than two days steaming distance from the grounds. The catcher boats are simple two-boat purse seiners of around 30 to 50 tons (60 to 70 feet in length). Some boats may be even smaller. The 'searcher' boats are really engaged in attracting rather than finding the fish, though they do look for indications of fish schools before they begin to use their lights. A complete fleet might consist of two transport vessels,

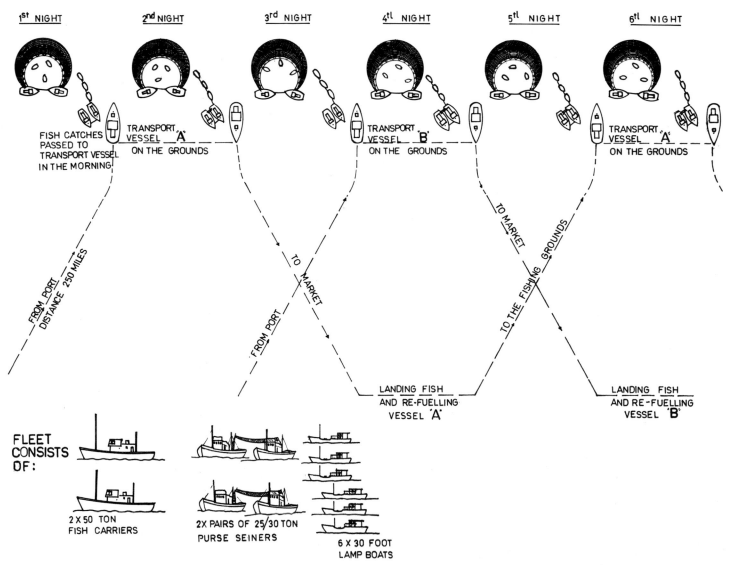

Fig 76   Fleet operation of two-boat purse seiners working with light attraction boats

four catcher boats and six light-boats. The light-searcher boats are small, around 30-35 feet in length with 20 or 30 hp engines.

A pair of 50-60 foot purse seiners might catch anything up to ten tons of fish in one night. Assuming an average nightly catch of five tons per pair, then the transport vessel could obtain ten tons each morning from two pairs.

The fish may be boxed in 15 kilo containers, with ice. On the second morning the transport vessel might leave for the market with 20 tons of boxed, iced fish aboard. Meanwhile, the second transport boat would be returning from port in time to take on board the following night's catch. On some fresh fish markets in South East Asia, such fish might sell for up to $1.00 (Singapore) per kilo (approx. $0.50 US). This would be equivalent to a British price of £198 a cran or £600 a ton. Thus at 20 tons a trip, each transporter making five trips a month, the little fleet could receive a reasonable financial return for their enterprise.

Successful fleet operations demand the same kind of team-work and trust between the skippers that is needed for two-boat fishing. This is probably why pair-fishermen are able to operate so well on a fleet basis. They have already developed the co-ordination and team spirit necessary. Fleet operations offer similar economic advantages for fishermen whose craft are too small or too lightly powered to engage in successful single-boat fishing, particularly when large distances are involved. The economics of each particular situation must be studied carefully to determine the most profitable method of operation.

Plate 46   Purse seiners assisting each other in British Columbia waters   *National Film Board*

Conversion Tables

Table 1

Fathoms to metres		Metres to fathoms	
0.5	0.914	0.5	0.273
1	1.829	1.0	0.547
2	3.658	2	1.094
3	5.486	3	1.640
4	7.315	4	2.187
5	9.164	5	2.734
6	10.97	6	3.281
7	12.80	7	3.828
8	14.63	8	4.374
9	16.46	9	4.921
10	18.29	10	5.47
25	45.72	15	8.20
50	91.44	25	13.67
75	137.2	50	27.34
100	182.9	75	41.01
125	228.6	100	54.7
150	274.3	125	68.4
175	320.1	150	82.0
200	365.8	175	95.7
225	411.5	200	109.4
250	457.2	250	136.7
275	503.0	300	164.0
300	548.6	350	191.3
350	640.0	400	218.7
400	731.5	450	246.0
450	823.0	500	273.4
500	914.4	750	410.1
600	1,097.3	1,000	546.8

Table 2

Feet and inches to centimetres		Centimetres to feet and inches	
⅛″	0.317	0.5	0.20″
¼″	0.635	1.0	0.39″
⅜″	0.952	2	0.79″
½″	1.270	3	1.18″
⅝″	1.59	4	1.57″
¾″	1.90	5	1.97″
⅞″	2.22	6	2.36″
1″	2.54	7	2.76″
2″	5.1	8	3.15″
3″	7.6	9	3.54″
4″	10.2	10	3.94″
5″	12.7	15	5.9 ″
6″	15.2	20	7.9 ″
7″	17.8	25	9.9 ″
8″	20.3	30	11.8 ″
9″	22.9	35	1′ 1.8 ″
10″	25.4	40	1′ 3.15″
11″	27.9	45	1′ 5.7 ″
		50	1′ 7.7 ″
1′ 0″	30.48	65	2′ 1.6 ″
1′ 3″	38.1	70	2′ 3.6 ″
1′ 6″	45.7	75	2′ 5.5 ″
1′ 9″	53.4	80	2′ 7.5 ″
2′ 0″	61.0	85	2′ 9.5 ″
3′ 0″	91.4	90	2′ 11.4 ″
4′ 0″	121.9	95	3′ 1.4 ″
5′ 0″	152.4	100	3′ 3.4 ″
6′ 0″	182.9	200	6′ 6.7 ″

Table 3
For Rope and Wire Sizes

Inches circumference to mm diameter		mm diameter to inches circumference		Inches circumference to mm diameter		mm diameter to inches circumference	
½	4.04	4	0.5	2¾	22.0	23	3.09
¾	6.06	5	0.62	3	24.3	30	3.71
1	8.08	6	0.74	3½	28.3	33	4.33
1¼	10.11	8	1.0	4	32.3	40	4.95
1½	12.13	10	1.24	4½	36.4	45	5.57
1¾	14.15	12	1.48	5	40.6	50	6.19
2	16.17	15	1.86	5½	44.5	55	6.80
2¼	18.19	18	2.23	6	48.5	60	7.42
2½	20.2	20	2.47				

Table 4

Nylon Twine Sizes (three strand, average construction)

Denier number	R Tex number
210/3	80
210/6	155
210/9	230
210/12	310
210/15	390
210/18	460
210/21	540
210/24	620
210/27	700
210/30	770
210/33	850
210/36	930
210/39	1,020
210/42	1,100
210/45	1,170
210/60	1,550
210/75	1,950

Table 6

Fish Room Capacities

Cubic feet to cubic metres		Cubic metres to cubic feet	
1	0.0283	1	35.32
10	0.283	2	70.63
20	0.566	3	105.95
25	0.707	4	141.26
30	0.85	5	176.6
40	1.13	6	211.9
50	1.42	7	247.2
60	1.7	8	282.5
70	2.0	9	317.8
75	2.12	10	353.2
80	2.26	15	529.7
90	2.55	20	706.3
100	2.83	25	882.9
125	3.54	30	1,059.5
150	4.25	40	1,412.6
175	4.95	50	1,765.8
200	5.66	60	2,119
250	7.08	70	2,472
500	14.15	75	2,649
750	21.23	80	2,825
1,000	28.3	90	3,178
1,500	42.5	100	3,532
2,000	56.6	125	4,415
3,000	84.9	150	5,298
4,000	113.2	175	6,181
5,000	141.5	200	7,063
6,000	169.8	300	10,595

Table 5

Mesh Sizes

Stretched mesh in inches to bar size in mm		Bar size in mm to stretched mesh in inches	
		5	0.4 "
½"	6.35	10	0.8 "
⅝"	8.0	15	1.2 "
¾"	9.5	20	1.6 "
⅞"	11.1	25	2.0 "
1 "	12.7	30	2.4 "
1⅛"	14.3	40	3.15"
1¼"	15.8	50	3.94"
1⅜"	17.4	60	4.72"
1½"	19.0	70	5.52"
1⅝"	20.7	75	5.92"
1¾"	22.2	80	6.3 "
1⅞"	23.8	90	7.1 "
2 "	25.4	100	7.9 "
2¼"	28.6	115	9.1 "
2½"	31.7	125	9.9 "
2¾"	34.9	135	10.7 "
3 "	38.1	140	11.0 "
3½"	44.4	150	11.8 "
4 "	50.8	175	13.8 "
4½"	57.1	200	15.75"
5 "	63.4	250	19.7 "
5½"	69.7	300	23.6 "
6 "	76.2	350	27.6 "
7 "	88.8	400	31.5 "
8 "	101.6	450	35.4 "
9 "	114.2	500	39.4 "
10 "	127.0	600	47.2 "
11 "	140.0	700	55.2 "
12 "	152.0	800	63.0 "

Note: It is common practice in some Scandinavian and European countries to express mesh sizes as bar size in millimetres. In Britain and the United States, stretched mesh size in inches is used. Stretched mesh size = 2 x bar size.

Table 7

Fish catch units and hold capacities

(Grimsby) 10 stone kits

Kits	= lb.	= kilograms
1	140	63.5
2	280	127.0
3	420	190.5
4	560	254.0
5	700	317.5
6	840	381.0
7	980	444.5
8	1,120	508.0
9	1,260	571.5
10	1,400	635.0
25	3,500	1,587.6
50	7,000	3,175

75	10,500	4,763
100	14,000	6,350
200	28,000	12,700
300	42,000	19,051
400	56,000	25,401
500	70,000	31,751
600	84,000	38,102
750	105,000	47,626
1,000	140,000	63,503

Herring units, old and new
Old crans = Unit crans

1	1.75	75	131.25
10	17.50	80	140.00
15	26.25	90	157.50
20	35.00	100	175.00
25	43.75	150	262.50
30	52.50	200	350.00
35	61.25	250	437.50
40	70.00	300	525.00
45	78.75	350	612.50
50	87.50	400	700.00
60	105.00	500	875.00
70	122.50	1,000	1,750.00

Notes on Stowage rates:
One cubic metre fishroom space should accommodate slightly more than:

    8 x 10 st. bulk kits and ice
    6 x 7 st. boxed and ice
    7 x unit crans, bulk
    ½ x tonne frozen fish

Note: 1 old cran = 37½ Imp. gallons.
    10 unit crans = 1 metric tonne.

Net Twine Sizes Used in Pair Trawls

All twine sizes for nylon except columns PE which indicate equivalent size for Polyethylene

Vessels	Overall length ft.	m.	Bottom Pair Trawls — Wings Rtex	Wings Denier	PE	Body Rtex	Body Denier	PE	Bag Rtex	Bag Denier	PE	Midwater Pair Trawls — Wings Rtex	Wings Denier	Body Rtex	Body Denier	Bag Rtex	Bag Denier
hp																	
2 x 5	20	6	310	210/12	380/9	310	210/12	380/9	310	210/12	380/9	230	210/9	230	210/9	230	210/9
2 x 10	25	7.5	390	15	12	390	15	12	310	12	9	390	15	310	12	310	12
2 x 25	30	9	460	18	15	390	15	12	390	15	12	460	18	310	12	310	12
2 x 50	35	10.5	620	24	18	540	21	15	460	18	15	540	21	390	15	390	15
2 x 75	40	12	700	27	21	620	24	18	540	21	15	620	24	460	18	460	18
2 x 100	50	15	770	30	24	620	24	18	540	21	15	700	27	540	21	460	18
2 x 150	55	16.5	990	39	27	770	30	24	620	24	18	850	33	620	24	540	21
2 x 200	60	18	1,250	48	36	990	39	27	770	30	24	930	36	700	27	620	24
2 x 300	65	20	1,550	60	42	1,250	48	36	990	39	27	1,100	42	770	30	700	27
2 x 400	70	21	1,700	66	48	1,550	60	42	1,250	48	36	1,250	48	850	33	770	30
2 x 500	75	23	1,900	72	54	1,700	66	48	1,550	60	42	1,400	54	990	39	930	36
2 x 600	80	24	2,300	96	60	1,900	72	54	1,700	66	48	1,550	60	1,250	48	1,100	42
2 x 750	87	26	2,800	108	72	2,300	96	60	1,900	72	54	1,900	72	1,550	60	1,250	48
2 x 1,000	100	30	3,500	135	96	2,800	108	72	2,300	96	60	2,300	96	1,900	72	1,550	60

Notes: In many cases, the lower wings of bottom trawls are made of heavier twine than the upper wings.
Double or heavier twine is used at fly meshes, seams and quarters.
In some nets the same size twine is used in wings, body and bag.
The bag part here indicated does not include the cod-end which is made of heavier twine.
These sizes are only approximate – for larger meshes, use heavier twine and vice versa.
In bottom trawls with only a gradual change in mesh size, the bag part is of heavier twine.

Details of Some Cod Pair Trawlers

Date	Builder	Name and Registry	Dimensions in feet	Main engine
1958	G. Thomson Buckie Scotland	*Solveig Borum* GY 508	65 x 17.5 x 7.5	Gardner 152 hp
1959	Herd & Mackenzie Buckie Scotland	*Athabasca* GY 293	73 x 20.4 x 9.5	Kelvin 240 hp
1958	G. Thomson Buckie Scotland	*Skanderborg* H 64	66 x 18 x 8	Gardner 152 hp
1973	Jones Buckie Scotland	*Frances Bojen* BCK 51	66 x 20 x 9	Gardner 230 hp
1975	Glyngore Denmark	*Carl Borum* GY 322	62.5 x 20.3 x 8.4	Alpha 400 hp
1974	Argyll Shipyard Scotland	*Mohave* GY 309	74 x 21.5 x 12	Blackstone 495 hp
1975	Glyngore Denmark	*Tino* GY 203	75 x 19.5 x 10	Grenaa 330 hp
1976	Jones Buckie Scotland	*Margrethe Bojen* BCK 111	68 x 20 x 10	Kelvin 280 hp

Note: All the above are traditional wooden vessels except the
Mohave which is a steel stern trawler.
Most of the vessels use a net drum, and a few also
have a power block.

Details of Some Scottish Pair Teams Using the Cod Trawl

Name	Number	Skipper	Type	Engine	Dimensions in feet
Fairweather V	PD 157	J. A. Buchan	Steel	637 hp	85 x 22 x 10
Sparkling Star	PD 137	J. Buchan	Steel	750 hp	87 x 24 x 12
Faithful II	PD 67	W. Milne	Steel	500 hp	79 x 22 x 10
Ugievale II	PD 105	A. Buchan	Wood	450 hp	78 x 21 x 10
Uberous	FR 50	I. Duthie	Wood	450 hp	78 x 22 x 9
Kallista	FR 107	G. Watt	Wood	850 hp	80 x 22 x 10
Seringa	PD 95	J. Morgan	Steel	600 hp	86 x 21 x 11
Sundari	PD 93	W. Morgan	Steel	600 hp	86 x 21 x 11
Unity	PD 209	J. W. McLean	Steel	660 hp	86 x 22 x 12
Morning Dawn	PD 195	D. Morgan	Steel	637 hp	86 x 22 x 12
Hesperus	BF 219	M. Watt	Steel	460 hp	75 x 20 x 10
Vesper	BF 220	G. West	Steel	460 hp	75 x 20 x 10
Amethyst	PD 74	J. Buchan	Steel	600 hp	85 x 22 x 12
Noronya	PD 168	A. Reid	Wood	495 hp	81 x 22 x 10

Details of Some Irish Midwater Pair Trawlers

Name and number	Builder	Type and dimensions in feet	Tonnage and capacity	Main engine	Deck equipment	Echo sounders	Home port
Bannrion na dTonn SO 444	Mevagh Yard, Ireland	Traditional 46.5 x 15.25 x 5.5	27 tons 100 crans	Gardner 110 hp	Sutherland winch	Simrad E/S	Donegal
Boy Evin D 560	Crosshaven, Ireland	Transom stern 52 x 17 x 7	38 tons 150 crans	Caterpillar 220 hp	Hydraulic winch And. & Sor. n.drum	Kelvin Hughes E/S	Clogerhead
Jemaleen SO 447	BIM, Killybegs, Ireland	Traditional 60 x 19.25 x 9.75	55 tons 200 crans	Caterpillar 240 hp	Mastra winch	Simrad E/S	Kincasslagh
Sea Bridger D 573	Tyrrell, Arklow, Ireland	Transom stern 65 x 19.66 x 8.5	70 tons 240 crans	Caterpillar 365 hp	Hydraulic winch	Simrad E/S	Killybegs
Fragrant Cloud B 224	Bangor Yard, Ireland	Traditional 65 x 20.5 x 9	82 tons 260 crans	Kelvin 415 hp	Hydraulic winch Lossie 19″ P.block	Kelvin Hughes E/S	Portavogie
Emily Francis WD 53	Maritem, Cork, Ireland	Traditional 66.6 x 20.5 x 8.8	80 tons 260 crans	Caterpillar 365 hp	Hydraulic Winch Hyd. net drum	2 x Simrad E/S	Kilmore Quay
Kenure D 359	BIM, Killybegs, Ireland	Traditional 70 x 20.4 x 10.9	104 tons 300 crans	Baudouin 430 hp	Bopp winch Ibercisa P.block	Simrad E/S	Skerries
Green Valley N 20	H. & M., Buckie, Scotland	Traditional 75 x 21 x 9	89 tons 320 crans	Caterpillar 425 hp	Norlau winch Rapp 19″ P.block	Kelvin Hughes E/S	Kilkeel
Angelique D 517	Herbert & Heraud, France	Traditional 75 x 21.3 x 11.2	105 tons 350 crans	Baudouin 360 hp	Robic & Corlay winch	Simrad E/S	Killybegs
Azure Sea D 585	BIM, Killybegs, Ireland	Traditional 78 x 21.25 x 11	110 tons 360 crans	Kelvin 500 hp	Brattvaag winch	2 x Simrad E/S Wesmar sonar	Galway
Fort Aengus G 4	BIM, Killybegs, Ireland	Traditional 80 x 22 x 12	120 tons 400 crans	Blackstone 600 hp	Robertson winch Carron 24″ P.block	Simrad E/S Simrad sonar	Kilronan
Carmarose SO 555	Iversen, Norway	Steel, transom st. 85.3 x 23 x 11.3	135 tons 500 crans	Caterpillar 850 hp	Lidan winch Rapp 24″ P.block	Simrad E/S Simrad sonar	Killybegs
Venture SO 523	Hakvoort, Holland	Steel stern trawler 86 x 24 x 11.5	145 tons 550 crans	Caterpillar 850 hp	Van de Geissen winch A.F. 19″ P.block	Furuno E/S Wesmar sonar	Killybegs

Notes: The term 'traditional' refers to the wooden cruiser-sterned designs with engine room and wheelhouse aft.
The two largest vessels listed have controllable pitch propellers and auxiliary engines of over 125 hp.
The smaller vessels carry a 15-35 hp auxiliary. Most of the boats have also a small landing winch.
Other standard electronic equipment not listed includes radio-telephone, Decca navigator, and radar.
Crans capacity are old crans, not unit crans.

Details of Some Scottish Midwater Pair Trawlers

Name and registered number	Builder	Dimensions in feet	Type	Fish hold capacity cu. ft.	Main engine
Crimond II KY 246	J. Noble, Fraserburgh	54.7 x 18.3 x 8	Wood, transom stern w.h. aft	1,400	Gardner 230 hp
Jacquamar FR 95	Forbes, Sandhaven, Fraserburgh	54.5 x 18.6 x 7.8	Wood, transom stern w.h. aft	1,400	Kelvin 280 hp
Providence IV LH 168	Mackay, Arbroath	55.5 x 19.6 x 8	Wood, transom stern w.h. aft	1,500	Kelvin 280 hp
Helenus UL 33	Herd & Mackenzie, Buckie, Banffshire	56 x 18.5 x 7.7	Steel, stern trawler	1,400	Caterpillar 300 hp
Silver Spray BCK 87	Herd & Mackenzie, Buckie, Banffshire	56 x 18.5 x 8.2	Wood, transom stern w.h. aft	1,500	Volvo 300 hp
Defiant OB 207	G. Thomson, Buckie, Banffshire	65 x 22 x 9.75	Wood, transom stern w.h. aft, engine forward	2,600	Kelvin 415 hp
Pathfinder FR 172	J. Noble, Fraserburgh	66 x 20.8 x 10.5	Wood, transom stern w.h. aft	2,200	Two Gardner 230 hp engines, one g.box
Stardust II LH 153	Mackay, Arbroath	66.7 x 21 x 12.5	Wood, transom stern w.h. aft	2,300	Caterpillar 380 hp
Athena LK 237	Jones, Buckie, Banffshire	72.3 x 22.5 x 14	Wood, transom stern w.h. aft	2,800	Kelvin 450 hp
Langdale PD 135	Bideford Yard, England	74.7 x 21 x 10.5	Steel, cruiser stern w.h. aft	3,280	Kelvin 500 hp
Falcon INS 117	Campbeltown, Argyll	75 x 21 x 10.5	Steel, cruiser stern w.h. aft	3,000	Caterpillar 480 hp
Hesperus BF 219	Lewis, Aberdeen, Scotland	75.5 x 20.4 x 10.5	Steel, transom stern w.h. aft	2,650	Deutz 460 hp
Uberous FR 50	Forbes, Sandhaven, Fraserburgh	78.5 x 22.2 x 9	Wood, cruiser stern w.h. aft	3,600	Caterpillar 450 hp
Christmas Rose FR 125	Campbeltown, Argyll	80 x 22 x 11	Steel, cruiser stern w.h. aft	4,100	Grenaa 500 hp
Noronya PD 168	Jones, Buckie, Banffshire	81 x 22 x 10.5	Wood, transom stern w.h. aft	4,000	Blackstone 495 hp
Amethyst PD 74	Southern Shipbuilders, Kent	84.8 x 22 x 12	Steel, cruiser stern w.h. aft	4,500	Blackstone 600 hp
Ajax INS 168	Campbeltown, Argyll	85 x 23.5 x 12	Steel, cruiser stern w.h. aft	5,400	Caterpillar 565 hp
Sparkling Star PD 137	Hakvoort, Holland	87.2 x 24 x 12.2	Steel, transom stern w.h. aft	5,500	Blackstone 750 hp

Auxiliary	Winch	Power block	Other equipment	Echo-sounder	Sonar	Other fishing methods used
	Mastra	Carron	Lossie rope reels	Kelvin Hughes		Seine net bottom trawl
Petter	Mastra		Netzonde cable winch	Elac		Bottom trawl
	Mastra	Lossie 19″	Beccles coiler rope bins	Koden		Seine net bottom trawl
Petter	Sutherland	Rapp 19″		Atlas		Bottom trawl
	Norlau	Rapp 19″	Seine rope reels	Kelvin Hughes		Bottom trawl seine net
Lister	Norwinch	A.F. 26″	Landing winch	Simrad	Simrad	Bottom trawl
	Mastra	A.F. 19″	Beccles coiler rope bins	Kelvin Hughes	Wesmar	Seine net
Petter 20 hp	Sutherland	Rapp 24″	Beccles coiler rope bins	Atlas		Seine net bottom trawl
Lister 44 hp	Sutherland	Carron 24″	Beccles coiler rope bins	Atlas		Seine net
Lister 44 hp	Jensen	Lossie 28″		Simrad		Bottom trawl
Lister 59 hp	Sutherland	Carron 24″	Beccles coiler rope bins	Kelvin Hughes		Seine net
Two Lister 32 hp	Dauntless	Lossie 28″	Beccles coiler rope bins	Atlas	Elac	Seine net
Lister 19 hp	Mastra	Rapp 19″	Beccles coiler rope bins	Elac		Seine net
Lister 22 hp	Jensen	Ibercisa	Beccles coiler rope bins	Kelvin Hughes		Seine net
Lister 37 hp	Jensen	A.F. 26″	Rope coiler rope bins	Kelvin Hughes		Seine net
Lister 32 hp	Robertson	Duplex X-2	Beccles coiler rope bins	Furuno	Furuno	Seine net
Gardner 110 hp	Mastra	Ibercisa 25″	FH rope reels	Elac		Seine net
Gardner 110 hp	Robertson	Rapp 24″	Robt. net drum	Elac	Elac	Purse seine

Details of Some Scottish Ring Net Vessels

Date	Builder	Name and Registry	Dimensions in feet	Main engine
1961	Jas. Noble, Fraserburgh	*Aspire* INS 148	55 x 17.5 x 7.0	Gardner 152 hp
1961	Alex. Noble, Girvan	*Village Maid* TT 25	58.5 x 18.3 x 7.5	Gardner 152 hp
1962	Alex. Noble, Girvan	*Excellent III* OB 89	45 x 16 x 6.0	Gardner 110 hp
1963	Herd & Mackenzie, Buckie	*Falcon* INS 235	60 x 18 x 8.7	Gardner 200 hp
1963	Alex. Noble, Girvan	*Crystal Sea* OB 104	60 x 18.3 x 9.0	Kelvin 240 hp
1964	Alex. Noble, Girvan	*Aliped IX* BA 234	59 x 18 x 7.0	Kelvin 180 hp
1964	W'head & Blackie, Port Seton	*Maryeared* TT 57	52 x 17.8 x 5.8	Gardner 200 hp
1964	Alex. Noble, Girvan	*Pathfinder* BA 252	59 x 18 x 7.5	Gardner 200 hp
1964	Forbes, Sandhaven, Fraserburgh	*Britannia* BA 267	59 x 18 x 7.5	Caterpillar 220 hp
1966	Herd & Mackenzie, Buckie	*Fair Morn* BA 295	61.5 x 18.3 x 9.0	Caterpillar 270 hp
1966	Herd & Mackenzie, Buckie	*Flourish II* INS 123	61.5 x 18.3 x 9.0	Gardner 200 hp
1969	W'head & Blackie, Port Seton	*Zephyr* INS 6	60 x 19 x 7.0	Gardner 150 hp
1969	Jas. Miller, St. Monance	*Scalpay Isle III* SY 19	60 x 19 x 8.5	Kelvin 240 hp
1971	Forbes, Sandhaven, Fraserburgh	*Heather Bloom* INS 110	65 x 20 x 9.5	Caterpillar 380 hp
1972	Alex. Noble, Girvan	*Prospector* BA 25	55 x 18 x 7.0	Caterpillar 300 hp

Notes: All of these boats were built of wood, and all but one had traditional cruiser sterns.
Most of them were equipped to engage in trawling, seining or scallop dragging in the off season.

Landings by Ring Net Vessels, Scotland 1961-1975

Ref.: Dept. Agriculture & Fisheries for Scotland.

Year	Catch (Imp. Tons)	Value (£ Sterling)
1961	30,500	641,000
1962	37,800	824,000
1963	55,800	677,000
1964	49,700	845,000
1965	37,000	1,022,000
1966	37,000	861,000
1967	31,300	772,000
1968	32,400	809,000
1969	38,000	1,044,000
1970	31,300	1,089,000
1971	26,200	1,036,000
1972	20,600	965,000
1973	12,100	805,000
1974	7,200	755,000
1975	2,400	211,000

Note: In 1969 there were 53 registered ring net vessels; in 1973 there were only 21. This would give the following average annual catch per pair of vessels:

Approximate Annual Catch per Pair

1969	1,440 tons	£40,000
1973	1,150 tons	£76,000

Two-Boat Midwater Trawlers: Scotland

Catches landed 1972-1975

Weight in Metric Tonnes

Year	Herring	Mackerel	Sprats	Other	Total
1972	76,189	563	38,348	3,983	119,083
1973	75,044	3,822	59,521	10,406	148,793
1974	72,533	5,147	55,536	732	133,948
1975	53,734	4,415	21,966	360	80,475
1976	44,206	6,956	27,773	1,398	80,333

Value in £ Sterling

Year	Herring	Mackerel	Sprats	Other	Total
1972	2,620,494	18,057	648,996	58,238	3,345,785
1973	4,450,658	193,514	1,589,851	408,826	6,642,849
1974	6,837,881	199,654	1,713,674	42,318	8,793,527
1975	5,156,386	199,071	610,837	26,015	5,992,309
1976	5,669,240	390,702	1,079,131	82,127	7,221,200

Details of Voyages

Year	Number of Trips	Number of Days	Average length of trip (days)	Average landing (metric tonne)	Average trip earnings per pair £ Sterling
1972	11,780	12,729	1.08	10.1	568
1973	13,987	15,631	1.17	10.6	950
1974	13,386	18,186	1.4	10.0	1,314
1975	8,992	12,753	1.4	9.0	1,332
1976	8,939	12,204	1.36	9.0	1,622

Source of figures: *Scottish Sea Fisheries Statistical Tables*

Over one hundred vessels are listed as using mainly the midwater pair trawl in Scotland. To this figure must be added a sizeable number of bottom trawlers and seine-netters who also use the gear for shorter periods.

Note: These figures do not include the lucrative fisheries at the Isle of Man, off North Shields and off the Cornish coast where Scottish vessels also operate.

Catches and Trip Grossings by Some Cod Pair Trawlers Operating from Grimsby 1976

Month	Vessels	Skipper	Catch/days	Gross £	Daily average £
January	*Frances Bojen* Skanderborg	Jens Bojen P. Scott	700 kits 12 days	16,570	1,380
January	*Frances Bojen* Skanderborg	Jens Bojen P. Pulfrey	698 kits 12 days	15,180	1,265
February	*Frances Bojen* Skanderborg	Jens Bojen P. Pulfrey	826 kits 11 days	15,500	1,410
February	*Sonia Jane* Anna Micelle	D. Sorensen M. Josefsen	694 kits 10 days	13,340	1,334
March	*Frances Bojen* Margrethe Bojen	J. Richardson Jens Bojen	1,177 kits 11 days	20,408	1,855
April	*Margrethe Bojen* Frances Bojen	Jens Bojen J. Richardson	1,149 kits 10 days	24,733	2,473
April	*Shawnee* Mohave	D. Brown C. Spall	854 kits 15 days	16,786	1,119
April	*Laurids Skomager* Paul Antony	Jorgen Bojen F. Josefsen	761 kits 12 days	17,444	1,453
April	*Margrethe Bojen* Frances Bojen	Jens Bojen J. Richardson	895 kits 11 days	17,204	1,564
May	*Mohave* Shawnee	C. Spall D. Brown	1,207 kits 14 days	23,393	1,670
May	*Margrethe Bojen* Frances Bojen	Jens Bojen J. Richardson	1,225 kits 11 days	27,739	2,521
May	*Skanderborg* Golden Venture	P. Scott P. Pulfrey	737 kits 10 days	14,895	1,489
May	*Shawnee* Mohave	D. Brown C. Spall	1,440 kits 14 days	29,896	2,135
May	*Laurids Skomager* Ann Charlotte	Jorgen Bojen R. Collins	700 kits 8 days	15,116	1,889
June	*Margrethe Bojen* Frances Bojen	Jens Bojen J. Richardson	940 kits 10 days	18,182	1,818
June	*Golden Venture* Skanderborg	P. Pulfrey P. Scott	963 kits 14 days	18,706	1,336
June	*Shawnee* Mohave	D. Brown C. Spall	1,436 kits 13 days	27,764	2,135
June	*Carl Borum* Jacqueline Borum	R. Borum J. Borum	998 kits 15 days	19,545	1,303
June	*Searcher* East Bank	B. Nejrup J. Lee	739 kits 9 days	17,563	1,951
June	*Grenaa Star* Grenaa Pearl	B. Host M. Potterton	844 kits 12 days	16,908	1,409
June	*Ann Charlotte* Laurids Skomager	R. Collins Jorgen Bojen	950 kits 12 days	19,284	1,607
June	*Ellen* Melissa Louise	A. Thinnesen M. Clark	721 kits 10 days	16,112	1,611

Month	Vessels	Skipper	Catch/days	Gross £	Daily average £
June	*Margrethe Bojen*	Jens Bojen	1,158 kits	24,550	2,231
	Frances Bojen	J. Richardson	11 days		
June	*Golden Venture*	P. Pulfrey	929 kits	19,116	1,593
	Skanderborg	P. Scott	12 days		
June	*East Bank*	J. Lee	659 kits	13,363	1,336
	Searcher	B. Nejrup	10 days		
July	*Carl Borum*	R. Borum	1,217 kits	19,022	1,188
	Jacqueline Borum	J. Borum	16 days		
July	*Margrethe Bojen*	Jens Bojen	1,223 kits	22,867	2,286
	Frances Bojen	J. Richardson	10 days		
July	*Grenaa Star*	B. Host	893 kits	17,237	1,436
	Grenaa Pearl	M. Potterton	12 days		
July	*Golden Venture*	P. Pulfrey	1,102 kits	18,194	2,021
	Skanderborg	P. Scott	9 days		
July	*Laurids Skomager*	Jorgen Bojen	942 kits	16,846	1,403
	Ann Charlotte	R. Collins	12 days		
July	*Searcher*	B. Nejrup	718 kits	16,976	2,122
	East Bank	J. Lee	8 days		
July	*Sonia Jane*	D. Bewley	849 kits	19,172	1,597
	Anna Michelle	M. Josefsen	12 days		
July	*Clee*	M. Cox	623 kits	13,978	1,397
	Martin Norman	D. Cox	10 days		
July	*Carl Borum*	R. Borum	807 kits	17,325	1,925
	Jacqueline Borum	J. Borum	9 days		
July	*Margrethe Bojen*	Jens Bojen	1,435 kits	33,461	3,041
	Frances Bojen	J. Richardson	11 days		
July	*Solveig Borum*	D. Sherriff	641 kits	14,405	1,309
	Ling Bank	D. Rose	11 days		
July	*Samantha*	H. Thinnesen	1,047 kits	22,099	2,009
	Tino	P. Thinnesen	11 days		
July	*Golden Venture*	P. Pulfrey	1,060 kits	24,031	1,848
	Skanderborg	P. Scott	13 days		
July	*Sonia Jane*	D. Bewley	883 kits	22,079	1,840
	Anna Michelle	M. Josefsen	12 days		
July	*Searcher*	B. Nejrup	896 kits	21,421	3,060
	East Bank	J. Lee	7 days		
July	*Martin Norman*	D. Cox	739 kits	19,331	1,610
	Clee	M. Cox	12 days		
July	*Laurids Skomager*	J. M. Call	932 kits	21,734	1,671
	Ann Charlotte	R. Collins	13 days		
July	*Margrethe Bojen*	Jens Bojen	1,147 kits	27,192	2,266
	Frances Bojen	J. Richardson	12 days		
August	*Shawnee*	D. Brown	843 kits	22,634	1,414
	Mohave	B. Almond	16 days		
August	*Samantha*	H. Thinnesen	618 kits	19,510	1,500
	Tino	P. Thinnesen	13 days		
August	*Golden Venture*	P. Pulfrey	783 kits	19,457	1,768
	Skanderborg	P. Scott	11 days		
August	*Grenaa Star*	B. Host	690 kits	17,985	1,798
	Grenaa Pearl	M. Potterton	10 days		
August	*Margrethe Bojen*	Jens Bojen	1,219 kits	29,095	2,909
	Frances Bojen	J. Richardson	10 days		
August	*Searcher*	B. Nejrup	682 kits	19,913	1,422
	East Bank	J. Lee	14 days		

Month	Vessels	Skipper	Catch/days	Gross £	Daily average £
August	*Martin Norman*	D. Cox	755 kits	15,955	1,227
	Clee	M. Cox	13 days		
August	*Paul Antony*	J. Zeebroek Sr.	893 kits	19,597	1,507
	Glenda	F. Wintein	13 days		
August	*Grenaa Star*	B. Host	827 kits	18,995	1,461
	Grenaa Pearl	M. Potterton	13 days		
August	*Margrethe Bojen*	Jens Bojen	1,350 kits	32,387	4,048
	Frances Bojen	J. Richardson	8 days		
August	*Laurids Skomager*	Jorgen Bojen	833 kits	20,632	2,292
	Ann Charlotte	R. Collins	9 days		
September	*Carl Borum*	R. Borum	963 kits	23,272	1,939
	Jacqueline Borum	J. Borum	12 days		
September	*Margrethe Bojen*	Jens Bojen	1,354 kits	36,342	3,303
	Frances Bojen	J. Richardson	11 days		
September	*Searcher*	B. Nejrup	923 kits	28,729	2,394
	East Bank	T. Trunbull	12 days		
September	*Anna Michelle*	M. Josefsen	721 kits	23,765	2,376
	Sonia Jane	D. Bewley	10 days		
September	*Golden Venture*	P. Pulfrey	665 kits	21,637	2,404
	Skanderborg	D. Auckland	9 days		
September	*Carl Borum*	R. Borum	996 kits	27,838	2,141
	Jacqueline Borum	J. Borum	13 days		
September	*Laurids Skomager*	J. Bojen	843 kits	24,990	1,922
	Ann Charlotte	R. Collins	13 days		
September	*Margrethe Bojen*	Jens Bojen	1,431 kits	40,979	3,152
	Frances Bojen	J. Richardson	13 days		
September	*Searcher*	B. Nejrup	584 kits	17,143	1,558
	East Bank	J. Lee	11 days		
October	*Margrethe Bojen*	Jens Bojen	782 kits	19,446	3,241
	Frances Bojen	J. Richardson	6 days		
October	*Golden Venture*	P. Pulfrey	996 kits	29,299	2,092
	Skanderborg	P. Scott	14 days		
October	*Margrethe Bojen*	Jens Bojen	554 kits	18,374	2,296
	Frances Bojen	J. Richardson	8 days		
October	*Samantha*	H. Thinnesen	743 kits	22,147	1,703
	Tino	P. Thinnesen	13 days		
October	*Anna Michelle*	M. Josefsen	656 kits	21,789	1,980
	Sonia Jane	D. Bewley	11 days		
October	*Mohave*	C. Spall	671 kits	20,701	1,881
	Shawnee	D. Brown	11 days		
November	*Margrethe Bojen*	Jens Bojen	854 kits	25,323	2,110
	Frances Bojen	J. Richardson	12 days		

Glossary

During his years in Africa and Asia, the writer has had many students ask him for an explanation of fishery terms which just do not translate literally, 'cod end' being a prime example. The following list of words and terms is provided to help non-English fishery students understand some of the technical jargon and colloquialisms in British and American fisheries. The identity of some fish species is indicated as common names are sometimes misleading.

Abeam　On a line at right angles to a vessel's length.

Aboard　On board (on a vessel).

Aft　At or towards the stern of a vessel.

Anchor seiner　Danish seiner (vessel bottom seining with an anchor).

Anchovy　Any of the *Engraulidae* family of species. These fish are small, but can be very plentiful as off Peru. They are used to make fish meal, fish paste and dried or salt fish.

Appearance　A sign or indication of fish presence, like a flock of sea birds for instance.

Athwartships　Along the beam of a vessel. At right angles to the keel or length.

Bacalao　Spanish for 'cod'. Sometimes used for salt cod fish.

Bar　One side of a net mesh, from knot to knot. A cut of 'all bars' on sheet netting is a straight diagonal cut.

Batings　Old name for the tapered part of a trawl net bag. Usually referred to the upper part.

Beam　The width of a vessel.

Becket　A rope or wire with loops in each end, used to fasten around the bag or cod end for lifting by block and tackle.

Belly　The under part of a trawl net bag. Being in close proximity to the sea bed, it often is torn and needs to be repaired or replaced.

Bight　Loop of rope.

Block　Pulley mounted to mast or derrick for lifting ropes to travel through. May be made of wood, steel or synthetic material with hard wearing pin and sheave.

Blue whiting　*Micromesistius poutassou*. Deep water species found in the east Atlantic from Ireland to Norway.

Boards　Abbreviation of otter boards.

Bobbins　Rollers strung on to trawl net groundropes to enable them to come over rough grounds. They may be made of wood, rubber, iron or plastic.

Bosom　The lower part of a trawl net mouth comprised of the middle of the footrope and the first part of the belly netting, usually reinforced.

Bow　Of a vessel, the part near the stem where the bulwarks flare out.

Box　Fish boxes come in many sizes, varying from 20 to 60 kilos. In Britain most fish boxes contain 7 stones (98 lbs) but some contain 8 stones (112 lbs) some 6 stones (84 lbs) and some 25 kilograms, of fish.

Box-trawl　Trawl net made up from 4 seams or panels.

Brailer　The long narrow meshed bag of a herring trawl. Also used of the brail net or dip net used to empty a purse seine.

Bridles　The ropes immediately connecting to the wing of the net. They may be double or triple on each side.

Bulk　Fish packed in holds without boxes or shelves are called bulkfish.

Bull-nose bobbin　Semi-spherical bobbin placed at the lower wing end or lower bridle end on a trawl net.

Bull trawl　Japanese two-boat trawl.

Butterfly　Steel 'boomerang'-shaped spreader between the bridles and groundcable of a trawl. Usually attached to a bull-nose bobbin.

Cables　Groundcables are trawl wires placed between the otter boards and the bridles or between warp end weights and bridles. They may be made of wire or of combination wire and rope.

Capelin　*Mallotus villosus*. Small smelt-type fish found in the north Atlantic from Newfoundland to Norway. Spawns on beaches.

Chafing gear　Any kind of protection for the underside of a cod-end. Cow hides were formerly used a lot. Synthetic chafers are now common.

Choker A splitting strap or becket around the forward part of a herring bag. Used to prevent fish from escaping and to hold the upper end of the bag when it sinks with the weight of a heavy catch.

Clip-links Chain connectors which are split and openable. May be used to connect two lengths of rope or warp.

Coalfish Saithe, *Pollachius virens,* one of the *Gadidae* species.

Cod *Gadus morhua* principal demersal fish of the north Atlantic.

Cod-end The final part of the trawl net bag where the fish end up. It is made of heavy or double twine to withstand chafing and to lift the catch on board.

Combination rope Heavy duty rope made of fibres and steel wire.

Cran Measure of herring quantity. The old cran was 37½ Imperial gallons of 4 quarter cran baskets, and this weighed about 400 lbs. The new unit cran is a measure of 100 kilograms weight of herring.

Crow's nest Look-out place on a ship's mast.

CRT Cathode Ray Tube scale expansion fish finder. The echoes appear as blobs on a central beam of light on the CRT screen.

Cruiser stern Sometimes called cruiser-spoon. Round-shaped stern on a 'double-ender' vessel, as opposed to a flat transom stern.

Cutter German, Polish, Danish or Dutch term for inshore trawler.

Cutting rate When cutting net panels from a sheet of webbing, the taper is determined by a 'cutting rate' formula of points and bars referring to the cut at each knot.

Dab *Limanda limanda*, common north Atlantic flatfish.

Dan leno Trawl bridle pole or spreader comprising butterfly and bull-nose bobbin.

Davits The light gallows used for lifting lifeboats on deck.

Decca navigator Electronic aid to navigation based on hyperbolic position lines obtained from signals by a series of master and slave stations ashore.

Deep sea trawler The term refers to larger long-distance craft which use heavy bottom trawls. It does not mean that they necessarily work in deeper water.

Dhan Marker buoy with a flag pole attached.

Dodge To steam slowly into strong winds or heavy seas, just fast enough to maintain steerage without making any real headway.

Dog ears American term for fly meshes.

Dog rope See **'Lazy line'.**

Doors Otter doors, or otter boards.

Drifter Vessel which uses drifting gill nets to catch fish, usually herring.

Fischelupe or fish scope, CRT display fish finder.

Flapper Loose hanging piece of netting in a trawl net bag, which acts as a kind of 'non-return valve', preventing fish from swimming out.

Fly mesh Square meshes on the outer edges of trawl net wings. With the replacement of hand-braided net with sheet cut netting, fly meshes are being replaced with cuts of all bars or drop-meshes, duly reinforced to strengthen them.

Fo'c'sle Literally 'fore castle', the forward accommodation on a vessel.

Gallon Liquid capacity measure. The British or Imperial gallon = 4.545 litres, the U.S. gallon = 3.785 litres.

Gallows The inverted 'U'-shaped frames which support the trawl towing blocks and up to which the otter boards or warp weights are hauled.

Gantry Strong steel 'goalpost'-shaped frame which on a stern trawler acts as lifting and towing point in lieu of gallows and derrick.

Gilson A single purchase wire rope with hook attached for lifting trawl gear.

G-links 'G'-shaped chain links with the aperture shaped to link up with each other; used to join ropes, bridles or lazy lines.

Grassrope Chafing rope made of coconut fibre. It is usually strung with lead weights and attached to the footrope by nylon cords.

Groundrope The trawl net wire on which the bobbins are strung. (Not to be confused with the footrope to which the netting is attached. Small or lighter trawls have a grassrope or chain instead.)

Haddock *Gadus aeglefinus,* demersal fish of the *Gaidae* family, next in importance to cod in the north Atlantic.

Hake *Merluccius merluccius* in the north east Atlantic. Other *Merluccius* species occur in the west, mid, and south Atlantic and SE Pacific. Red Hake in the west Atlantic, *Urophycis chuss,* are quite different.

Halfer Fisherman's term for a 3 leg knot or mesh knot with one bar broken.

Hanging ratio	The ratio of the length of the headline or footrope to the length of the netting to be attached. In times past the rope was always shorter than the netting. In modern synthetic nets, the reverse may be the case.	**Knocking-out**	The operation of releasing trawl warps from the towing block on a side trawler, is called 'knocking-out'. It is dangerous especially in bad weather as the release of tension can throw the block inboard. Modern hydraulically released blocks are far safer than the old chain-connected pin-gate blocks.
Headline height	The vertical mouth opening of a trawl net in operation.		
Hemp	A tough bast fibre obtained from the Asiatic plant of the same name. It is not much used in fishing gear now.	**Lastridge**	On a trawl net where two panels are joined, the side seam contains a few meshes bunched together for strength. They are laced rather than joined mesh-to-mesh, hence 'lastridge' or selvedge.
Herring	*Clupea harengus,* the most important pelagic fish in the north Atlantic. Many other clupeoid species occur in different seas. The herring is remarkable for its food value and is marketed in a multitude of different ways.	**Lay**	Can refer to the share-out system of crew remuneration. Of a rope it refers to the direction of twist (right-hand or left-hand).
Horse mackerel	Also called jack mackerel; any of the *Trachurus* or *Decapterus* species. Those mentioned in the text also include *Rastrelliger* and *Selaroides* species.	**Lazy line**	or lazy decky; a strong rope from the choker or cod-end which is tied loosely to the headline. It is used to pull in the bag via the gipsy drum until the fish are alongside.
Independent wire	or Pennant wire: a short piece of wire connected to the warps at either side of an otter board or weight but not taking any strain during the tow. It is used to disconnect the otter board and continue hauling the cables after the warps are all in.	**Lengthener**	Extension piece on a trawl net bag.
		Light trawl	A high-opening bottom trawl of light construction. They were a Scandinavian invention, there called 'vinge' or wing trawls.
Jigging	Fishing with hooks having artificial lures. The lines are tugged or 'jigged' and fish are attracted by the movement of the lure.	**Live-bait**	Small fish kept alive in sea water tanks for use as bait to attract skipjack and bonito.
Kelly's Eye and keep	The ring and figure-of-eight-shaped shackles which allow the pennant wire to draw the cables to the winch when hauling, but which will jam at the pennant when the cables are paid out.	**Loran**	(from LOng RANge) Radio navigation system involving signals from pairs of stations. Used extensively by fishing vessels in the north Atlantic, it has greater range though less accuracy than the Decca system.
Kippering	A kipper is a herring that has been split and smoked. The kippering trade is an important part of the herring industry. Unfortunately many of today's kippers are dyed and not smoked.	**Lozenge trawl**	Four-seam midwater trawl design in which the seams run from the middle of the head, foot and side ropes instead of from the wing end points as in a Larsen trawl.
Kit	An old English measure for trawl-caught fish, its weight may be different in different ports. The Hull, Grimsby, Fleetwood kit contains 10 stones (140 lbs or 63.5 kilos). The Granton and Aberdeen kits contain 8 stones (112 lbs or 50.8 kilos).	**Mackerel**	Any of the *Scombridae* family, Atlantic mackerel are *Scomber scombrus*. All mackerel are fast swimming pelagic fish with no air bladder.
		Manila	Vegetable fibre from Asia used to make good quality ropes. The manila plant is related to hemp.
Klondyking	This refers to the old practice of salting and icing herring quickly for transport to a distant centre where they would be further processed. The full method of gutting and salting herring was much more time-consuming and labour intensive.	**Markings**	Echo sounder recordings of fish schools.
		Mask	Scottish or Danish for 'mesh'.
		Menhaden	*Brevoortia* species. 'Pogies', found off south and mid-eastern USA, oily bony fish used almost wholly for industrial purposes.

Mesh	The basic unit of a sheet of netting, its size is usually determined by measuring the stretched distance from one knot to the next diagonally opposite (stretched mesh). Bar size is measured from knot to knot across one leg of twine. See also under **Points.**
Messenger	After shooting warps on a side trawler, the forward gallows warp is pulled in to the towing block aft by means of a messenger wire and hook. The messenger hook has a very sharp bend to it.
Monofilament	A twine constructed of a single thick fibre of synthetic material.
Multifilament	A twine constructed of many thin fibres of synthetic material.
Mullet	Any of the *Mugilidae* species found chiefly around the north Atlantic, they are not plentiful although prized as table fish.
Net sounder	or Netzonde, a transducer fixed to the headline of a trawl and connected by cable to an echo sounder on the vessel to determine the net depth and mouth opening.
Ossels	On a light trawl or drift net, the cords connecting the grassrope or mud rope to the foot. They are made of braided nylon or spun manila twine.
Otter board	The heavy horizontal spreaders which give otter trawls their name.
Overhang	At the mouth of a trawl net, the upper sheet usually 'overhangs' the lower bosom by the length of its 'square' net panel. The horizontal distance it stretches forward is called the overhang.
Panel	The netting pieces making up a trawl are called panels.
Pilchard	Generally refers to the larger sardines. The English pilchard, *Sardina pilchardus* is an olive-coloured fish closely resembling a herring. The famous South African pilchard is *Sardinops ocellata*.
Points	In net cutting a 'point' is a side knot as opposed to 'meshes' which are mesh loops above or below a sheet of netting. A point cut is a vertical cut and a mesh cut is a horizontal cut when the netting is hung by the length *ie* so that the continuous twines run horizontally.
Pole & line	The gear used to catch skipjack in the tropics. It usually does not carry any bait. For big fish one line may be attached to two poles.
Pollock	*Pollachius pollachius*, found in the NE Atlantic. Related species are found in the west Atlantic and in the north Pacific (*Theragra chalcogrammus*).
Polyamide	The technical name for 'nylon' synthetic fibre.
Polyester	Synthetic fibre known by trade names 'Terylene', 'Dacron', 'Tetoron', *etc.*
Polyethylene	Synthetic fibre known by trade names 'Courlene', 'Nymplex', 'Hi-Zex', *etc.*
Polypropylene	Synthetic fibre known by trade names 'Ulstron', 'Nufil', 'Pro-Zex' *etc.*
Polyvinyl alcohol	Synthetic fibre known by trade names 'Kuralon', 'Vinylon', *etc.*
Port	On a vessel, looking forward, the port side is the left side. See **Starboard**.
Pounds	The compartments for fish on deck and in fish holds, formed by loose wooden boards slipped into slotted pillar supports.
Power block	Vee-shaped powered roller for hauling nets. It is powered hydraulically and its size is measured by the width across the outer end of the 'V'-shaped sheave.
Prawns	In Europe this refers to Norway lobster (*Nephrops norvegicus*) which are marketed as 'scampi'. Unlike shrimp, they have claws.
Purser	Any vessel using a purse seine.
Purse rings	The brass or stainless steel rings through which the pursing line runs around the foot of a purse seine.
Quarter	On a vessel, the part aft where the hull curves towards the stern.
Quarters	On a trawl net, the corners between the wings and the square or bosom. They are usually strengthened by double twine.
Quarter ropes	Lifting ropes which are attached to the bobbins or groundrope at the lower quarters, and run through a becket at the upper quarters. They are loosely tied to the wing ends. When the net is alongside they are bent on to the winch to lift the heavy bobbins on board.
Rigging	Of a net, the act of fitting it with floats, weights, shackles, *etc.*
Ringer	Any vessel using a ring net.
Sardine	Any of the *Sardinella* species. Also used of pilchards and small herrings, the term may be erroneously applied to sprat or brisling.
Seam	In a net, where two panels join lengthwise. Most bottom trawls are 2 seam nets, most midwater trawls are 4 seam nets.

162

Selvedge   Usually on the side seams of a trawl net, the selvedge is composed of several meshes from each panel, lashed together with twine which is hitched at intervals rather than at every turn. The selvedge adds strength to the seam. The word can also mean reinforcing or heavy twine bordering any type of net.

Shelving   The method of packing fish in shelves fitted in the hold lockers. The shelves reduce the pressure that would occur with bulk packed fish. However, although they look better in appearance, shelf fish are reckoned to be of poorer quality than bulk packed fish.

Shingle   On the sea bed, heavy gravel mingled with broken shells.

Shoal   1. A school of fish.
2. A shallow bank under the sea; shallow water.

Shoe   Of an otter board, the heavy iron base which slides over the sea bed.

Shoot   To pay out the gear while the vessel makes way through the water.

Shoulder   Of a trawl net, the first part of the wings where they join to the square and belly.

Side knot   In net mending, going from left to right and right to left, the knot at which one changes direction and commences another row. A 'point' in net cutting.

Sisal   Vegetable fibre grown chiefly in Mexico and Latin America. It produces a rope similar to manila but of less strength and durability.

Skiff   Strictly speaking the term refers to a small flat-bottomed boat but it is now used to refer to purse seine assisting boats which are designed to act like miniature tugs.

Skipjack   *Euthynnus pelamis*, sometimes called Bonito, a surface swimming fish of the tuna family. It weighs 2 or 3 kilos or more and feeds voraciously on small fishes swimming near the surface.

Snap-on   Connecting clips which can be snapped open and closed are snap-on clips. They are used chiefly on long lines. Snap-open rings are purse rings that operate in a similar manner.

Sole   1. General term for flatfish of the *Solea* species, the name is also used for several other flatfish.
2. The term may be used of the bottom or foot of a net or the floor of a fish hold.

Splitting rope   or splitting strap, the rope around the cod-end of a trawl net which 'splits' or separates the catch when more fish is in the bag than can be lifted aboard at one time.

Spot   On an echo recording it refers to a mark of a dense school of fish.

Sprat   *Sprattus sprattus*, a small pelagic fish found in dense shoals in the east Atlantic around the British Isles and Scandinavia. It is marketed as brisling in Norway.

Spring rope   The foot hauling rope on a ring net.

Square   The upper panel of a trawl net that overhangs the net mouth. On the early beam trawls it was square in shape but it is now much wider with the sides tapering from the leading edge.

Staple fibre   A synthetic twine, like nylon, can be spun from fibres than have been deliberately cut short. The result is a twine or rope with good gripping qualities. That is a staple fibre twine (sometimes called spun nylon).

Starboard   On a vessel, looking forward, the starboard side is the right side. (Originally the steering board was on this side leaving the other side clear to come against the pier or 'port').

Stem   The forward part of a vessel where it cuts the water.

Stern   The after or rear end of a vessel.

Take-up   or take-in; when two sheets of netting are joined together, and they do not have an equal number of meshes at the join, the difference is the take-up. For instance, if the end of the 'square' panel had 200 meshes and the first upper belly sheet had 240 meshes, the take-up rate would be one extra mesh every five.

Taper   The net panels on a trawl taper towards the wing ends or towards the bag. The degree of taper required was achieved by batings and creasings on hand-made nets. Modern sheet-cut panels are tapered by 'cutting rates' based on points and bars formula.

Tex   The Tex numbering system for net twine yarns is based on their mass in grams per 1,000 metres. R tex applies to completed twines and expresses their mass in grams per 1,000 metres.

Three-leg   See **Halfer**.

Tickler chain   To stir the sea bed in front of a trawl net and force flatfish to swim upwards, a chain is sometimes rigged between the wing ends, and of a length to permit it to drag a foot or two in front of the bosom. That is a tickler chain. Sometimes several chains are used.

Tom weight   A weight slid down purse seine wires to prevent them from rising and fouling the net while it is being pursed.

Towing block   The snatch block which holds the trawl warps on a side trawler during a tow. It is attached to the vessel by a short length of chain.

Towing roller   Small trawlers and bottom seiners use a roller fitted into the rail instead of a towing block.

Troller   Not to be confused with trawler. A troller catches fish by means of trolling lines extending from booms on either side. The method is used a lot in the Pacific for skipjack and yellowfin tuna, and off British Columbia and Alaska for salmon.

Tuna   Any of the large *Thunnidae* family. Sometimes called 'Tunny'.

Unit cran   Measure of weight for herring, equal to 100 kilos.

Vee doors   Otter boards having an angled rather than a flat surface. They are made of steel with a single swivelling bracket. In operation they are very stable and can pass over small obstacles easily.

Vee wing   High opening trawls now generally have the wing ends tapered in towards the central selvedge or seam. It is a typical feature of wing trawls. The Vee rope can be adjusted to transmit as much tension as required to the selvedge.

Vinge trawl   Scandinavian name for a wing trawl. The G, is pronounced hard.

Vigneron Dahl gear   or VD Gear, the name being taken from its French innovators, refers to the use of cables between the otter board and the net. More specifically it can refer to the 'Kelly's Eye' and 'Keep' which are an integral part of the gear on traditional vessels.

Warp   The wire or rope used to tow a trawl net.

Warp marks   Most trawlers mark their warps every 25 fathoms to determine the length paid out when shooting, and to keep both warps even. Sometimes a 1, 2, 3, 4, mark system is used to indicate the length more clearly.

Whiting   *Gadus merlangus* in the east Atlantic. In the west Atlantic the name is used of Silver Hake, *Merluccius bilinearis,* but the fish are quite different. The stocks of whiting are now greatly depleted in some areas, like the Irish Sea.

Wings   Of a trawl net, the two parts extending from either side of the trawl mouth like two arms or leaders to guide the fish in.

Yellowfin   *Thunnus albacares*, one of the tuna family. Larger than skipjack though not as big as bluefish tuna, it is found near the surface.

Zipper line   Rope passed through a series of rings to close off sections of a purse seine when separating large catches. Zippers may also be fitted to midwater trawls, but in that case to open the bag so the catch may be brailed out instead of using the cod-end and splitter system.

Bibliography

Account of the Fishing Gear of England and Wales	Davis
Canadian Pair Seining Experiment 1969	Rycroft
Catalogue of Fishing Gear Designs, Volumes 1, 2 and 3	FAO
Commercial Fisheries Review, Dept. Interior	USA
Commercial Fishing journal, Fleetwood	Britain
Fanggerate der Kutter – und Kustenfischerei	Von Brandt
Fischfangtechnik Fangtechnologie, Berufsschullehrbuch	Germany
Fisheries of the North Pacific	Browning
Fishing Boats of the World	
Volume 1 Japanese Drag Net Boats	Takagi
,, Dutch Coastal Fishing Boats	Zwolsman
,, Spanish Fishing Vessels	Caruncho
,, Discussion – Cutters	Hansen, Lembke
Volume 2 Purse Seining: Deck design and equipment	Schmidt
,, The Netherlands Post-War Fishing Fleet	Boogard
Fishing News (weekly), London	Britain
Fishing News International, London	Britain
France Peche, journal	France
How To Make And Set Nets	Garner
Illustration of Japanese Fishing Boats, Fisheries Agency	Japan
Illustration of Japanese Fishing Nets, Fisheries Agency	Japan
Irish Skipper, journal, Irish Maritime Press, Dublin	Ireland
Korean Fishing Gear, Fisheries Research and Development	Korea
La Peche Maritime, journal	France
Le Chalut	Nedelec, Libert
Midwater Trawling (IFGC Newfoundland)	Schärfe
Midwater Trawling (National Fisherman) 1975	Innes
Midwater Trawling, Reports WFA	Britain
Midwater Trawls, Akrehamn Trälbøteri	Norway
Modern Fishing Gear of the World	
Volume 1 Studies on Two-Boat Trawls	Hamuro, Ishii
,, German Cutter Trawling Gear	Schärfe
,, Midwater Trawls and their Operation	Parrish
,, Scandinavian Experience Midwater Trawling	Larsson
,, The Thames Floating Sprat Trawl	Noel
,, Menhaden Purse Seining	Robas
,, The Puretic Power Block	Schmidt
Volume 2 Towing Power, Speed and Size of Bull Trawl	Hamuro
,, Two-Boat Midwater Trawling for Herring	Steinberg
Volume 3 Purse Seining and Aimed Trawling	
,, History of Purse Seining in Japan	Inoue
,, The Purse Seine Fishery in Japan	Akaoka
,, Japanese Tuna Purse Seining	Sugano, Yamamura
,, Two-Boat Bottom and Midwater Trawling	Steinberg
,, New Dutch Two-Boat Midwater Trawling	Minnee
More Scottish Fishing Craft and their Work	Wilson
National Fisherman, journal, Camden, Maine	USA
Netting Brochures, Nitto Seimo Co. Ltd.	Japan
Netting Brochures, Nam Yang Co. Ltd.	Korea

Nets, Twines and Ropes, Fukui Fishing Nets	Japan
Offshore Fishing, journal, London	Britain
Pair Trawling, Scottish Fisheries Bulletin	Corrigall
Pair Trawling, World Fishing 1971	Corrigall
Preliminary Study of Trawl Net In Thailand, IPFC	Nomura
Progress Report: Canadian Pair Seining	King
Scottish Fishing Craft	Wilson
Small Boat Pair Midwater Trawling, Dept. Fisheries, Canada	Brothers
Scottish Fisheries Bulletins, Dept. AFF	Scotland
Scottish Sea Fisheries Statistics, HMSO	Scotland
Sea Fisheries Statistics (England and Wales)	HMSO
Sea Food Ships	Hardy
Textbook, Coastal Fisheries of Japan, OTCA	Japan
Trawling and Continuous Fishing, USSR	Kostyunin, Nikonorov
Two-Boat Midwater Trawling in New England, NEMRIP	MacLeod, Taber
Two-Boat Ring Net Fishing for Capelin, Report	Canada
World Fishing, journal, London	Britain
Yearbook of Fishery Statistics	FAO

List of other books published by
Fishing News Books Limited

1 Long Garden Walk, Farnham, Surrey, England

Free catalogue available on request

A living from lobsters
Advances in aquaculture (proceedings of FAO conference)
Aquaculture practices in Taiwan
Better angling with simple science
British freshwater fishes
Coastal aquaculture in the Indo-Pacific region
Commercial fishing methods
Control of fish quality
Culture of bivalve molluscs
Eel capture, culture, processing and marketing
Eel culture
European inland water fish: a multilingual catalogue
FAO catalogue of fishing gear designs
FAO catalogue of small scale fishing gear
FAO investigates ferro-cement fishing craft
Farming the edge of the sea
Fish and shellfish farming in coastal waters
Fish catching methods of the world
Fish farming international No. 2
Fish inspection and quality control
Fisheries oceanography
Fishery products
Fishing boats and their equipment
Fishing boats of the world 1
Fishing boats of the world 2
Fishing boats of the world 3
Fishing ports and markets
Fishing with electricity
Fishing with light
Freezing and irradiation of fish
Handbook of trout and salmon diseases
Handy medical guide for seafarers
How to make and set nets
Inshore fishing: its skills, risks, rewards
International regulation of marine fisheries: a study of regional fisheries organisations

Marine pollution and sea life
Mechanisation of small fishing craft
Mending of fishing nets
Modern deep sea trawling gear
Modern fishing gear of the world 1
Modern fishing gear of the world 2
Modern fishing gear of the world 3
Modern inshore fishing gear
More Scottish fishing craft and their work
Multilingual dictionary of fish and fish products
Navigation primer for fishermen
Netting materials for fishing gear
Pelagic and semi-pelagic trawling gear
Planning aquaculture development
Power transmission and automation for ships and submersibles
Refrigeration on fishing vessels
Salmon fisheries of Scotland
Salmon and trout farming in Norway
Seafood fishing for amateur and professional
Ships' gear 66
Sonar in fisheries: a forward look
Stability and trim of fishing vessels
Testing the freshness of frozen fish
Textbook of fish culture; breeding and cultivation of fish
The fertile sea
The fish resources of the ocean
The fishing cadet's handbook
The lemon sole
The marketing of shellfish
The seine net: its origin, evolution and use
The stern trawler
The stocks of whales
Trawlermen's handbook
Tuna: distribution and migration
Underwater observation using sonar

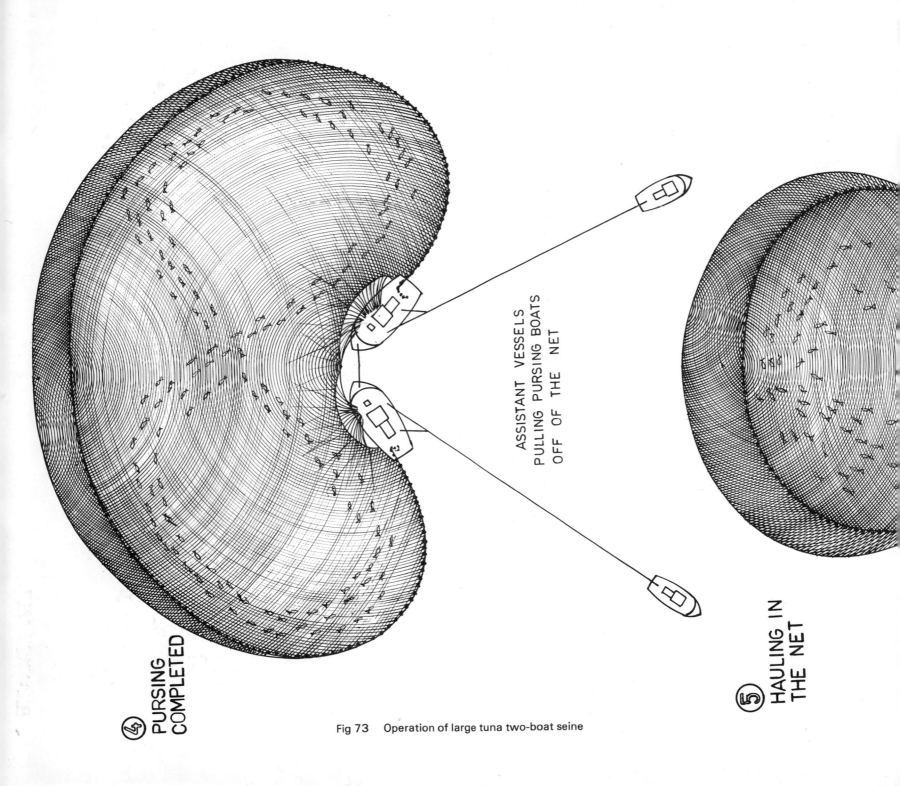

④ PURSING COMPLETED

ASSISTANT VESSELS
PULLING PURSING BOATS
OFF OF THE NET

⑤ HAULING IN THE NET

Fig 73   Operation of large tuna two-boat seine